Life on Man

Also by Theodor Rosebury

Experimental Air-Borne Infection

Peace or Pestilence

Microorganisms Indigenous to Man

Life on Man

THEODOR ROSEBURY

NEW YORK / THE VIKING PRESS

First published in 1969 by The Viking Press, Inc.
625 Madison Avenue, New York, N.Y. 10022

Published simultaneously in Canada by
The Macmillan Company of Canada Limited

Library of Congress catalog card number: 69-18804

Printed in U.S.A.

ACKNOWLEDGMENTS

George Allen & Unwin Ltd: From *Introductory Lectures to Psychoanalysis* by Sigmund Freud.

Basic Books, Inc., Publishers: From *The Collected Papers of Sigmund Freud*, edited by Ernest Jones, M.D. (1959). Reprinted by permission.

Edmund R. Brill: Quotations on page 64 are from *The Basic Writings of Sigmund Freud*, translated and edited by Dr. A. A. Brill, Copyright 1938 by Random House, Inc. Copyright renewed 1965 by Gioia B. Bernheim and Edmund R. Brill. Reprinted by permission.

Dover Publications, Inc.: From *Antony Van Leeuwenhoek and His Little Animals* by C. Dobell.

Sigmund Freud Copyrights Limited, The Institute of Psycho-Analysis and the Hogarth Press Ltd., for permission to quote from the *Standard Edition of the Complete Psychological Works of Sigmund Freud*.

Little, Brown and Company and J. M. Dent & Sons Ltd.: From *Verses from 1929 On* by Ogden Nash. Copyright 1953 by Ogden Nash. Reprinted by permission of Little, Brown and Company and J. M. Dent & Sons Ltd.

Liveright, Publishers: From *A General Introduction to Psychoanalysis* by Sigmund Freud. Copyright 1920, 1935 by Edward L. Bernays.

W. H. Lowdermilk & Co.: From *Scatological Rites of All Nations* by J. C. Bourke, W. H. Lowdermilk & Co. (1891).

Professor Horace Miner: From *Body Ritual Among the Nacirema*. Reprinted by permission of author.

Routledge & Kegan Paul Ltd: From *Lars Porsena* by Robert Graves. Reprinted by permission.

FOR CELIA

whose name derives from Shakespeare rather than from Swift or that other, obscurer poet—although, to be sure, their Celias were all admirable girls, too.

. . . if you know whether you should show it to Ladies? Yea in any wise to all maner of Ladies, of the Court, of the country, of the City, great Ladies, lesser Ladies, learned ignorant, wise simple, fowle well favoured, (painted unpainted) so they be Ladies, you may boldly prefer it to them . . . Howbeit, you must not show it after one fashion to all, but to the wise and sober after a plaine fashion; to the wanton and waggish, after another fashion; as namely, if they cry (fie, for shame) when they hear the title read, or such like; do but you say (for company) that it is a mad fantastickal booke indeed, and when you have done hide it away, but where they may finde it, and by the next day, they wilbe as cunning in it as you . . .

. . . I would keepe it, as I would from fire and water . . . First, from a passing proud fellow . . . Secondly, from all manner of fooles and jesters . . . Thirdly, if you spie a fellow with a bay leafe in his mouth, avoyd him, for he carieth a thing about him worse than Maister A Jax, that all the devises he have cannot reform . . . and Fourthly, if you see a stale leane hungrie poore beggarly thridbare Kavalliero . . . show it not to him, for though he can say nothing against it, yet he will leere under his hat, as though he could speake more than he thinkes. For such a one that mak's not a goode meal at home once in a month, hath not a good stoole above once in a weeke . . .

—Sir John Harington, *The Metamorphosis of Ajax*

Author's Note

This book is intended mainly for readers without special scientific or scholarly pretensions, but scholars and scientists are invited to read it as well, and an effort has been made to please all. For the sake of the general reader, superscripts have been omitted in the text (except one).* But notes and a bibliography are supplied; the former are keyed forward to text pages and to the words or phrases that apply, and, where appropriate, backward to the bibliography. A preliminary glance at the back of the book will show what has been done. The reader who must have superscripts can easily supply them with his own hand.

---

* *An asterisk in the text and a pointer footnote on the same page mark the first note to appear.*

Contents

Preview

Shame take him that shame thinks.

—Old English proverb

The color and odor of the fluid suggest cleanliness, and the bottle is pretty but austere, like a starched nurse. The voice on TV says it kills germs, kills them by the million.

This is today's mythology. Cleanliness has moved up from being merely next to godliness into a religion in itself. We are becoming a nation of tubbed, scrubbed, deodorized neurotics. Once it was only pleasure that was dirty and sinful. Now there is dirt everywhere, and there are germs to prove it. How irrational can we get?

If you are healthy and your teeth are clean, unless you have been eating onions, your mouth doesn't smell. If it does you should see a dentist. Perfectly healthy young adult mouths contain germs *by the billion*—which means, of course, by the thousand million. No mouth is without them. Even if something really did kill them by the million it would be doing only one-thousandth of the job. But as it kills a few germs it also damages the cells of your mouth and interferes with other things provided by nature that need no help in keeping your mouth healthy—including some of the microbes, which destroy other microbes. Not only are you wasting your money on this beautiful rubbish, but it would not be worth using if you got it for nothing. It does harm without doing good. At best the harm is not noticed, and you may settle for a clean sensation. You may mask an odor, but if it didn't come from something you ate, the odor should be treated, not masked.

The same goes for other body odors. Blood and sweat do not smell until they are decomposed. Up to a point there is nothing unhealthy

xiii

about the decomposition, even though microbes bring it about. But the use of a deodorant within bounds is a cosmetic act, like the use of lipstick and perfume. Only when it is done excessively, obsessively, does it become neurotic. The use of antiseptics, however, can hardly ever be justified. The attempt to kill the microbes that live on us normally is a mistake.

We have become too fastidious. Our noses have become so absurdly sensitive that we wrinkle them at the slightest hint of a healthy smell, which is not too strong a term for the smell of sweat from hard work or active play. We don't object to these smells in ourselves or in people we love unless they are forced on our attention. In learning to find them intolerable we are aping the dandified, exquisite aristocrat of the seventeenth century in his powdered wig, his laces and satins. There wasn't much plumbing then either for him or for the sweaty peasant. He shared the peasant's lice and fleas as well as his smells. But he masked his superior status by wearing ornate dress and strong perfume. We do things differently now, pretending to have the support of science. But our excesses are no more scientific than his were, and they are even further from common sense, since we ought to know better.

Most body odors are produced by microbes; and although we objected to some of them long before microbes were discovered, we have come to associate the two things. Germs produce disease, and so we think of them as nasty little things in their own right. Traces remain of the puritan notion that our bodies, or parts of them, their functions and products, are ugly, dirty. Sex has been entangled in this notion. Feces, filth, dirt, soil, earth—these are all related. Now that we know that germs are active in all of them, it looks as though there may be sense in the whole idea. But microbiologically speaking—and for other good reasons as well—it just isn't so.

Some of the other good reasons have been contributed by Freud rather than by Pasteur. Some of them have come from anthropologists and archeologists. And, in fact, some of the most telling reasons have been given to us by artists, poets.

Freud has made us see, for instance, that our reaction to excrement tends to go far beyond the bounds of rationality. We all share in some degree the delusion of the "anal" person who keeps washing his hands and picking threads off his clothing, trying desperately to hide from the world an unconscious notion of inescapable filthiness. We are all in some degree compelled toward a spotlessness we can never achieve in fancy or in fact, nor would there be any sense in it if we could.

The anthropologists tell us that primitive man tended—just as infants do today—to venerate his body and to cherish what it produced. It is characteristic of Western civilized man that he has become extraordinarily anal in Freud's sense. Young people today, including many college students, often prefer to emulate the more alienated hippies rather than their elders, rejecting many of our values, including the whole business of neatness. They see cleanliness as part of the sham of a hypocritical world. They are also forcing a sexual revolution upon us; and we are inclined to sympathize with part of it, since the irrationality of regarding sex as something dirty is becoming inescapable. Yet the dimensions of our irrationality extend well beyond sex—which is certainly part of what poets and artists have been trying to tell us since long before Rabelais, before Chaucer, since the times of Aristophanes and Praxiteles and Sappho, of Solomon's Song of Songs.

We are starting to teach the biology of reproduction to children and hope for rationality and sanity in generations to come. Perhaps we ought to do the same with the microbiology of man, for the same purpose of encouraging rational behavior. We ought to reconcile the science of epidemic disease with that of soil fertility, and show that the microbes living normally on man are a kind of bridge between these two, an aspect of the over-all scheme of living things in the world. These microbes fall neatly between the "evil" of disease and the "good" of the earth's fertility. They are closer to the second, since in the normal course they do us no harm. In fact we know now, from experiments with animals we can make and keep free from germs—they turn out to be miserable, deprived things—that we could not get along without microbes.

"Dirt," considered as "earth," is not evil. The myth that germs and dirt are always our enemies is harmful and costly. We ought to get rid of it. The evil of disease remains, of course. We need to put it in perspective. Disease lends itself to measurement and analysis by the methods of science. It has shown itself to be amenable to understanding, treatment, control, even the possibility of eradication. For our purposes we can point out the connection between disease and cleanliness and set limits to it. It can be made clear that too much cleanliness is as sick as too little.

This book is about the microbes that live on man, what they do and what they mean to us. Microbiology—or, as we sometimes say, bacteriology, meaning the same thing—is part of it, but only part. I intend to give you that part, in words you don't have to be a scientist to understand; and I will tell you how we get along with our microbes, and how they get along with us. But even more important, I think, is what this all means to us, what the whole notion of "dirt" and "cleanliness" has meant through the ages before the idea of microbes existed, what it means today. The microbes that live on us are part of the "dirt"—speaking broadly, the *excrement*—that we try hopelessly and irrationally to get rid of. We are going to look into the origins of this compulsion, in anthropology and in Freud, in history and in literature. Much of this discussion gets to be a long way from bacteriology, but bacteriology puts it in a special light when the whole subject is brought together. Or so it seems to me. And since I am dealing with a touchy subject and finding my way—and yours too—through some old prejudices and phobias, I propose to go cautiously, and even, when it seems to me desirable, circuitously or obliquely. If some of my digressions turn out to be interesting in themselves, or even amusing, so much the better.

PART I

Microbes Without Shame

1

Tom Jones's Mouth

To say the truth, as no known inhabitant of this globe is
really more than man, so none need be ashamed of submitting
to what the necessities of man demand.

—Henry Fielding, *Tom Jones*

Some day, the newspapers tell us, a man—or more likely
two or three—will step down on the surface of Mars. Perhaps they
will find evidence of something living there so clear and plain that
their television cameras will send pictures back to us that will settle
the question once and for all. If not, they will attempt to bring back
samples of whatever they find for analysis, and one of the things to
be looked for will be evidence of any life on the planet. We worry
about spoiling the experiment by having our Martian messenger contaminated with life from the earth. The contamination would be
microorganisms. If any signs of life are found, we'd like to be sure
it is native to Mars. We might be convinced of its Martian origin if
it turned out to be entirely different from anything we know on earth.
. But the more we learn of living things, the less probable this idea
seems. The organization of life in cellular and molecular terms is so
distinctive and essentially uniform on earth that it is no longer easy to
imagine—as it once was—that life elsewhere could be constituted on
a different basis. And astronomers and wise men in the budding field
called "bioastronautics" now seem to be agreed that life must exist
elsewhere in the universe—if not on Mars, certainly on planets like
Earth in other solar systems. There are so many star-planet systems
in the unimaginable vastness of space, they say, that somewhere there
must be a planet sufficiently Earthlike in history, composition, and
climate to support life. If the conditions that allow life to develop

3

exist, it is argued, given only enough time—millions and billions of years—it would be expected to develop. The cells of life and their component molecules would almost certainly be similar to those of life on Earth, the argument goes, because it is hard to imagine that so intricate and near-perfect a mechanism, so uniform on Earth despite its diversity of detail, is anything but inevitable wherever conditions allow it to begin and to grow.

But as the development proceeded we would expect small differences in the conditions to determine differences in the organisms that evolved. However many life-supporting planets there may be, we wouldn't expect any two of them to be identical. On our own planet, variations of climate in different places, and total isolation of one place from others for long periods, must account for the appearance of the kangaroo and other distinctive marsupials only in Australasia, and even for geographical variations in the single species *Homo sapiens*. Hence, while wise men assure us that there must be life in outer space, they give us little reason to believe that any sort of man will be found there. Nor can our vanity or our arrogance allow us to believe that outer space contains no organism older than ourselves, more highly developed, superior to us in any or every imaginable way—or, for that matter, even more aggressive, ferocious, and destructive than we are. Maybe there is a crumb of comfort in the notion that greater or older ferocity and destructiveness than ours would by now have burned the planet in question to a fission-fusion cinder. If we ourselves survive long enough, the most useful cosmic lesson in store for us could be the discovery in space of the burnt-out remains of what once must have been a planet more splendid than anything in our own great and beautiful and murderous past.

It is all fascinating, the more so because ideas like these have definitely moved out of the realm of science fiction, have passed from unlikely to possible to probable, as witness the billions being spent on space exploration, not to speak of the savants with their sophisticated electronic ears listening continuously for messages across the light years.

As we train our senses and focus our disciplined imaginations on the vastness of space it seems to me that we are missing something close to home. There is more than the contempt of familiarity in the extraordinary circumstance that, in our eagerness to learn of life far beyond earth, we overlook the life on our own surface—on the surface of man himself. Everybody is at least dimly aware that parts of our body are populated, and is likely to shudder at the idea. Even experts on microorganisms are usually a little too farsighted to be more than vaguely aware of those on themselves. The subject is unfashionable if not more or less repulsive. Whatever the reason, there is an almost universal ignorance, among specialists and educated people as well as others, of information as close to us and as open to view as our mouths. If scientists are repelled by the subject, others are apt to be revolted by it. But, with increasing knowledge, revulsion sometimes yields to fascination.

For several years I have been brooding about a way to bypass this difficulty, after devoting a good part of my working life to the scientific side of the subject. Several opportunities to lecture on this theme to technical groups looking for relaxation led me in 1964 to approach it in a way that may still be useful. At that time the wide-screen color movie *Tom Jones* was delighting people everywhere—myself among them. Like many others, I was stimulated by a particular sequence in the film, and I want to remind you of it, or to describe it in case you missed seeing it. Henry Fielding's original will help, although the particular scene I have in mind is one of several in which the screenplay, written by the notable British playwright John Osborne, and directed by Tony Richardson, had departed from the novel. In this instance the change, in addition to suiting my special purpose to perfection, seems to me to have preserved intact the spirit of the original and to have given it a twentieth-century flavor and a breathtaking presence as well. There were eloquent gasps in the audience as the scene came on! The Mad Hatter would have called it larger than life and at least twice as natural. It was almost micro-scopically real; and that's the point.

I may be able to bring the picture back to you, if you saw it, or to give you some idea of its splendor, if you didn't, with the help of Fielding's own words. The sequence comes almost precisely in the middle of the story. Tom has left Squire Allworthy's hearth and his beloved Sophia and is on his adventurous way to London. A fight with Ensign Northerton over Sophia's good name has nearly killed him, and the ensign is unaware that he has recovered. Tom has met Partridge, his devoted servant, but the intricacies of the plot require that the two be separated at this juncture. Tom then comes upon Northerton proceeding to hang a buxom and half-naked woman from the limb of a tree. He rescues the lady, who (although not known to the reader of the book) is recognizable to the viewer of the movie as

Jenny Jones, presumably Tom's mother. Jenny, having been driven from the Allworthy premises years before, does not recognize Tom but immediately accepts him as a good angel.

> Indeed he was a charming figure, and if a very fine person and a most comely set of features adorned with youth, health, strength, freshness of spirit, and good nature can make a man resemble an angel, he certainly had that resemblance.

The two then proceed toward Upton, silhouetted against the brow of a hill, Tom in the lead so as to shield his eyes from Jenny's nakedness,

> as Orpheus and Euridice marched heretofore, but though I cannot believe that Jones was designedly tempted by his fair one to look behind him, yet as she frequently wanted his assistance to help her over stiles and had, besides, many trips and other accidents, he was often obliged to turn about.

But Tom, of course, does not suffer the fate of Orpheus and comes safely with Jenny to the inn at Upton. In the original—somewhat modified in the movie—rollicking complications ensue: a free-for-all fight occasioned by Jenny's undress, with Partridge temporarily back in the picture as a belabored participant; the arrival at the inn in sequence of Sophia and her entourage, of her father, Squire Western, of Squire Allworthy himself, and of the sergeant who resolves the fracas by identifying Jenny as Mrs. Waters, the captain's wife, a lady despite her déshabille. These events intertwine so that the principals just fail to meet. Tom misses Sophia but finds her muff; she picks up his trail, thereby adding to his woes, and leaves before her father arrives; and Partridge, nursing a bloody nose at the pump, misses Jenny, whom he would have recognized. All this non-recognition resembles King Lear's initial stupidity in being essential for the unwinding of the plot.

Jenny's good character having been established and her bosom covered with borrowed raiment, she and Tom—in the movie—sit down to break a long and arduous fast, facing each other across a board groaning with dishes in spectacular color: this is the scene. Fielding treats it differently, pointing to its immediate sequel rather than to the meal itself, with Tom concentrating on three pounds of the flesh of an ox and Jenny eating little through preoccupation with other aims. As important to our story as the quotation at the head of this chapter (which appears at this point in the book) are these additional words of Fielding's:

Heroes, notwithstanding the high ideas which by the means of flatterers they may entertain of themselves or the world may conceive of them, have certainly more of mortal than divine about them. However elevated their minds may be, their bodies at least (which is much the major part of most) are liable to the worst infirmities and subject to the vilest offices of human nature.

But this suggestion of ugliness is quickly counteracted with another description of Tom, who

was in reality one of the handsomest young fellows in the world. His face, besides being the picture of health, had in it the most apparent marks of sweetness and good nature . . . He was, besides, active, genteel, gay, and good-humored, and had a flow of animal spirits which enlivened every conversation where he was present.

And so, by different means but to the same effect, we are treated to a scene in which Jenny, no shrinking violet, her heart captured by this paragon who, like a medieval knight, has rescued her from the ogre, levels upon him her "whole artillery of love." Repulsed more than once by intervention of "the God of Eating," in a battle reminiscent of the one between Hero and Leander with the sexes reversed, Jenny finally breaches Tom's "Dutch defense" (much as Marlowe said, in the instance of Hero, "In such battles women use but half their strength") and leads her lover to bed. Be it noted that the suggestion of incest, apparent at this point in the movie and emerging a good deal later in the book, is dispelled with pre-Victorian propriety when it appears, at the very end, that Jenny was not Tom's mother after all. In Fielding's day he had to be a gentleman. This is a long way from *Oedipus Rex*. It is even possible to imagine that in this context incest would not have mattered very much. It is part of Osborne's genius that the idea of incest could be suggested in the movie, and the bedroom scene shown—the latter being no more than delicately hinted at in the book—without changing the general spirit of the tale. Despite its lustiness, the movie has been generally accepted as just as wholesome as the much more sanitary *The Sound of Music*.

It is the eating scene in the movie that I have been leading up to. We see the two faces, Tom's and Jenny's, close up and in brilliantly real color, consuming, as I remember, a sequence of soup, lobster, chicken, mutton, oysters, and wine. I can't remember ever having seen anything like it before, in or out of real life. In the real world it is hard to imagine being so close, and yet no more than an onlooker, to the spectacle of two handsome people addressing themselves simultaneously with their fingers and their mouths to food one can all but savor and taste, and with their eyes to one another. The murmurs and gasps contributed by the audience testified to the force of the scene. I think the general effect was the one I felt myself—faithfully reflecting the spirit of Fielding, of robust, healthy enjoyment in the affirmation of the intense pleasure that life occasionally provides.

But to me there was something else. Modern zoom lenses and magnificent color reproduction with full movement can bring us so close to pulsating life that we see things no painter could imagine before these processes existed. The biologist—the oldfashioned kind who uses a light microscope—knows that while the picture changes as we move closer, as we bring a smaller and smaller area into focus, it loses none of its fascination, indeed none of its beauty, provided that we are prepared for what we find. This is one way in which we see the continuity of the living world, its never-ending wonder and magnificence, from the greatest to the smallest. And so, adding an imaginary zoom microscope to Richardson's color camera, I was able in my mind's eye to zero in on the little fleshy crevices around Tom's and Jenny's teeth as they ate their meal, and to see the turmoil of microbic life there, the spirochetes and vibrios in furious movement, the thicker corkscrew-like spirilla gliding back and forth, and the more sluggish or quiet chains and clusters and colonies of bacilli and cocci, massed around or boiling between detached epithelial scales and the fibers and debris of cells and food particles. Like the great and beautiful animals in whose mouths they live, these too are organisms, living things; and I could imagine them, quite like Tom and Jenny, making the most of the sudden accession of nourishment after a long fast.

2

Leeuwenhoek Saw It First

For my part I judge, from myself (howbeit I clean my mouth
like I've already said), that all the people living in our United
Netherlands are not as many as the living animals that I carry
in my own mouth this very day.

—Antony van Leeuwenhoek (1683)

$\mathbf{M}$aybe we had better slow up a little at this point. Widespread resistance to my subject is centuries old, and I can't hope to
dispel it with a kind word or two. If we are to get to the microscopic
center of this thing with our eyes open and our stomachs steady we
might do better to look gingerly and sip instead of gulping.

Something that happened to me many years ago just once—the
one experience warned me against repeating it—may help to explain
why I am being cautious if you don't otherwise know. As a long-time
teacher of bacteriology with a particular interest in the microbes on
the surface of man, I had the regular task of demonstrating them to
students. It happens that the best source of these microbes is the
mouth, especially that part of the mouth we were just speaking of,
the gum clefts at the bases of the teeth. The microbic population of
this region makes a particularly striking demonstration. This is far
from being the only populated region of our surface; but there is a
picturesque collection of microbes here, which has the additional
virtues of being easily accessible and not requiring undressing or
other embarrassments, to provide a sample for class use. So we look
for a suitable mouth. All human mouths except those that have lost
all their teeth have these microbes. For demonstration purposes,
to be sure of finding really enormous numbers of typical ones, we
look for areas of gum disease—so-called gingivitis or pyorrhea—

9

where overgrowths of the same microbes are easily found. Gingivitis is common enough in young adult students, but rather than look for it there I have usually found an older person whose value for the purpose could be spotted at a distance: a case of pyorrhea aggravated by neglect. The experienced hand then knows exactly where and how to gather the material and can guarantee a demonstration that is spectacular at first sight. I have seen countless numbers of such preparations, but the picture has never lost its fascination for me.

The experience I am recalling happened when I was a young instructor at the medical school of Columbia University, in a department with a large staff of teachers, assistants, and other people. On the occasion in question I chose one of the cleaning women as my subject, having noted previously that she had pyorrhea, mainly from the fact that her upper incisor teeth were splayed and protruding. She not only agreed willingly but was curious enough to want to look through the microscope; and, having set up the preparation, I let her be the first to look at it.

This was a mistake. Her reaction at seeing what she had in her mouth was instantaneous and obvious. She got away as soon as she could, and in the next few days she had a dentist extract all her teeth, and later appeared wearing a set of artificial ones. Even though this was doubtless the best thing she could have done for her mouth's sake, I didn't enjoy having been the tactless instrument of a kind of shock therapy that seems to me quite inappropriate in this application. Not only had I scared the wits out of the poor women, however unintentionally, but since the students, who crowded after her at the microscope and cut off her retreat, responded to the demonstration in a way that would otherwise have been gratifying, she must have felt as though she had been stripped naked before this group of young men, revealing to them monstrosities she had not previously suspected in herself. Or maybe she felt as any reasonably sensitive adult might feel if he were suddenly and uncontrollably incontinent before a mixed group of strangers.

These intense feelings of shame can't be overcome by denying them. Yet it is true that we can be thunderstruck to have something revealed about us that we know to be equally true of everyone else. We are not born with feelings like this. We learn them, and we ought to be able to unlearn them.

It seems to me extraordinary that the first person in history known to have seen these microbes of the mouth—or any microbes—showed no sign of revulsion toward them, although he recognized such feelings in others. He was Antony van Leeuwenhoek (pronounced *Lay-wen-hook*), a draper of Delft, who was born there in 1632 and died there in 1723—a span of 91 years. The painter Jan Vermeer was born in the same town in the same year, and died there too, but after a lifetime less than half as long, in 1677. What we know of Leeuwenhoek comes mainly from the labors of the eminent English biologist Clifford Dobell, who wrote a loving and scholarly book about the Dutchman, first published in 1932. Paul de Kruif, whom Dobell mentions in his dedication, had previously begun his *Microbe Hunters* with a chapter on Leeuwenhoek. Dutch scholars have not neglected their famous countryman; a book by A. Schierbeek tells us, among other useful things, that the compilation of the whole range of Leeuwenhoek's work is still going on.

Leeuwenhoek was not the first to make and use microscopes. Galileo and others even before him are said to have made them, but Leeuwenhoek's instruments seem to have been more powerful than the earlier ones. He became interested between 1660 and 1673, as Dobell says,

> —in his spare time, when he was not selling buttons and ribbon— in making lenses, and mounting them to form "microscopes" of simple pattern; and after he had acquired much skill in the manufacture of these curious instruments, and had taught himself how to grind and polish and mount lenses of considerable magnifying power, he began to examine all manner of things with their aid.

In 1673 a letter of Leeuwenhoek's describing some of his work was forwarded by the biologist Reinier de Graaf to the newly founded Royal Society of London, which then communicated with Leeuwenhoek and became the recipient of a series of letters from the microscopist himself, letters upon which his fame rests. Let me mention that England and Holland were at war at the time, and that Leeuwenhoek was an amateur scientist. Neither of these circumstances was remarkable. War did not then interfere in correspondence between scholars of belligerent countries, and scientists tended to be amateurs and were in no way the worse for being so.

His microscopes were "simple" in having only a single lens system, in contrast with modern "compound" microscopes, in which the image formed by one lens, the objective, is further magnified by another, the ocular or eyepiece. There are many other differences between Leeuwenhoek's instruments and the highly refined microscopes of today; yet the word "crude" is hardly appropriate for these early devices. We can only marvel at their relative perfection and at what Leeuwenhoek was able to see through them. Among his instruments that still exist, the strongest magnifies 270 times and has a resolving power of 1.4 microns, that is, it can show as two separate points or lines two objects 0.0014 millimeters apart. Today's relative figures are, for magnification, approximately 1500 diameters, and for resolving power, about 0.2 micron. This is for light in its visible range of wavelengths. Present-day instruments are limited not by human ingenuity but by the wavelength of light. Higher magnifications can be reached by using light of shorter wavelength—ultraviolet, which is invisible to the eye but not to the camera—or with an electron beam, which makes possible enormously greater powers, 100,000 diameters and more. These refined methods have the disadvantage of being usable only with fixed or immobilized cells; they cannot show us movement, as microscopes that use visible light can—both modern ones and Leeuwenhoek's.

The resolving power of light microscopes can be improved, and moving particles shown that would otherwise be unclear or invisible, by a trick of indirect lighting. If instead of shining the light through the specimen to the eye, as is usually done, we beam it from the side or at an angle, we see the object by light reflected off it to the eye, and it looks bright against a dark background. An old variant of this method was called the "ultramicroscope" in the days before electron microscopy, and was used to show the "Brownian" movement of submicroscopic particles responding to the random impact of invisible molecules around them. For biological work an instrument called the darkfield or dark-ground microscope is used. Dobell thinks Leeuwenhoek may have used such a method:

> All the evidence indicates that it was *the method of using* this apparatus which he "kept for himself alone": his secret lay, as he tells us repeatedly, in his "particular *method* of observing." What can it have been? The answer is—to me—almost certain, although I cannot prove from his own words (since he tried not to give the secret away) that I am right. I am convinced that Leeuwenhoek had, in the course of his experiments, hit upon some simple method of *dark-ground illumination*. He was well aware, as we know, of the ordinary properties of lenses; and he tells us himself that he used concave magnifying mirrors and employed artificial sources of illumination (e.g. a candle). Consequently he may

well have discovered by accident—or even have purposely de-
vised—some method which gave him a clear dark-ground image.
Such a discovery—possibly inspired by observing the motes in a
sunbeam—would at once explain all his otherwise inexplicable
observations, without supposing him to have possessed any apparatus other than that which we know he had. But no hint was
ever knowingly given, in all his many letters (so far as I have
been able to ascertain), of what his "particular method of observing" may really have been.

Techniques of some kind are probably always required for the
exercise of genius; but something else is needed, too. In Leeuwenhoek
the techniques must have been at hand, and a driving, uninhibited
curiosity was there; but beyond these admirable things he must also
have had pertinacity and perspicacity to an extraordinary degree. He
saw things nobody had ever seen before, described them accurately,
made sketches of them, and—most astonishing of all—estimated how
many of them there were in a unit of volume. He used his own unit,
a sand grain, which he was careful to define. His estimates have a
precision we can duplicate only now, with vastly better microscopes
and other devices and methods he never dreamed of.

As Dobell says, Leeuwenhoek examined "all manner of things." He
trained his lenses on the world around him, especially on animals
and plants, and on their smaller and smaller parts, and on himself.
He looked at everything with cool impartiality. His curiosity was all-
embracing. Wonder excited him at fresh discovery wherever he turned
his lenses, which showed him everywhere a world never seen by man
before. No fear, no revulsion deterred him. In this he was as innocent
as a child to whom a great and wonderful world slowly unfolds before
his developing senses, except that, as Leeuwenhoek had no fond but
fearful parent to tell him, "Be careful!" "Don't touch!" "Naughty!"
"*Shame!*" the inhibitions we think of as normal never appeared.

Very early in his letter-writing career—on October 19, 1674—we
find Leeuwenhoek describing what seem to have been protozoa in the
bile of animals, and thus discovering parasitic microbes, as he had
earlier found free-living ones. In a letter written on November 12,
1680, to his fellow microscopist Robert Hooke (whose lenses were
not powerful enough to show microorganisms), in which he described
the eggs and spermatozoa of insects, he also mentioned what seem to
have been protozoa in the gut of a horsefly. Then, a year later, on
November 4, 1681, he spoke of his own "excrements," again in a letter

to Hooke. It seems that a transient diarrhea had directed his attention to this material, and to some of his most exciting discoveries. In this letter he describes first the flagella-bearing protozoan of feces now known as *Giardia lamblia*:

> . . . animalcules a-moving very prettily; some of 'em a bit bigger, others a bit less, than a blood-globule; but all of one and the same make. Their bodies were somewhat longer than broad, and their belly, which was flatlike, furnisht with sundry little paws, wherwith they made such a stir in the clear medium and among the globules, that you might even fancy you saw a pissabed* running up against a wall; and albeit they made a quick motion with their paws, yet for all that they made but slow progress . . .

And then, a few lines later, with disarming casualness, he describes in the same material:

> . . . a sort of animalcules that had the figure of our river-eels: these were in very great plenty, and so small withal, that I deemed 500 or 600 of 'em laid out end to end would not reach to the length of a full-grown eel such as there are in vinegar. These had a very nimble motion, and bent their bodies serpentwise, and shot through the stuff as a pike does through the water.

Here we may stop to notice two things. First, the so-called vinegar-eel, of which Leeuwenhoek speaks repeatedly, is identified by Dobell as a tiny nematode worm, *Anguillula aceti*, a full-grown specimen of which is approximately 1.5 millimeters long. This would make the microbes in the quoted passage some 3 microns long, which is too short by a factor of about 3 or a little more: perhaps that is what the precise Leeuwenhoek meant by "would *not* reach." To the quoted passage Dobell appends this footnote:

> From the description here given, it can hardly be doubted that these organisms were spirochaetes. In recent years a vast literature has sprung up on the intestinal species in man; but it must suffice to note here that more than one species occurs in human faeces, and that spirochaetes of some sort are normally present in the intestines of most human beings. This is the first record of their occurrence.

As spirochetes are usually included among the bacteria (with some dispute from protozoologists that has, I think, been diminishing in

---

* If you think the word "pissabed" is worth stopping for, see the note at the end of the book.

recent years), this casual note of Leeuwenhoek's is also the first clear mention of parasitic bacteria, of bacteria "normal" to any animal, and hence of life on man.

But Leeuwenhoek could find little of interest, to him or to us, in his "excrement when it was of ordinary thickness," even though he diluted it with water; nor in the freshly collected "dirt of cows and horses, just as they let it drop." Yet in both feces and horse urine that looked abnormal to him, "thick, and . . . of an ashen color" he found

> . . . globules . . . which had a sixth of the bigness of one of our blood-globules, and also some whereof I judged that 36 of 'em together would only make up the bulk of a blood-globule.

These Dobell recognizes as bacteria, presumably rather large cocci, slightly more than 1 micron in diameter.

That Leeuwenhoek was unable to find much of interest in normal feces, which contains more microbes than does any other human or animal material, is less surprising than it at first seems to be. Most of the bacteria in feces are even smaller than the cocci just mentioned, and Leeuwenhoek may not have been able to distinguish them. The identification of the microbic content of feces has had to await methods other than microscopy, and is in fact still incomplete. Even today, Leeuwenhoek's having stumbled on *Giardia* and on *Borrelia*-like spirochetes would seem no more than a lucky accident. When Dobell speaks of a "vast literature" on intestinal microbes he cannot be referring to bacteria striking in appearance, such as spirochetes, on which the literature has always been small. But let us continue with Dobell:

> Before we proceed to the next discoveries, it may not be amiss to emphasize the novelty of those just recorded. At the time when the foregoing observations were made, no protozoa or bacteria of any kind were known—except the free-living forms described by Leeuwenhoek himself a few years earlier. No animal or "animalcule" smaller than a worm was known to live inside the body of man; and the existence of hordes of microorganisms within the bodies of healthy animals was as wholly unsuspected as it was unheard-of.

Dobell mentions that Leeuwenhoek did not think of associating these microbes with disease, although contemporary medical men quickly began to speculate on their pathological possibilities. Before Louis Pasteur and Robert Koch—two centuries later—such speculation could seldom be productive or convincing. What impresses me, again, is the absence in Leeuwenhoek of the morbid cast of mind that might have encouraged such ideas—to put the point more positively, his healthy, wholesome outlook.

During the period from 1674 to 1678 Leeuwenhoek made several references in his letters to "phlegm" and saliva, which are mentioned by Dobell in a footnote. These particular letters are incomplete, and the findings given in them seem unremarkable. In fact, in the famous letter of 1683 to which we are now coming, he says of saliva that "I could make out no animalcules there." But he goes on to examine material from around his teeth, with spectacular results. The letter is dated September 17, 1683, and is addressed to Francis Aston, secretary of the Royal Society. It includes a drawing that is probably as famous among bacteriologists as is Leonardo's "Mona Lisa." Let me quote several passages and comment on them as we go:

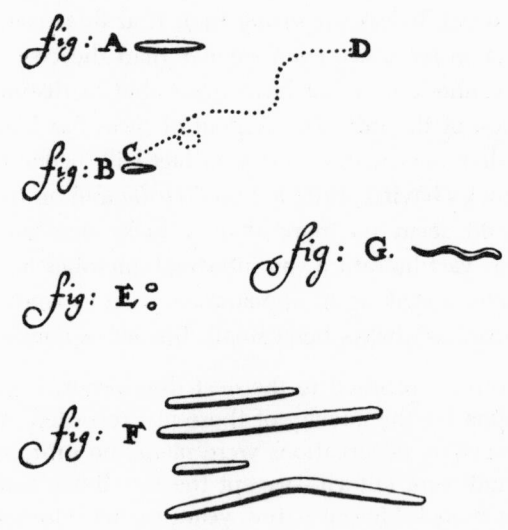

'Tis my wont of a morning to rub my teeth with salt, and then swill my mouth out with water: and often, after eating, to clean my back teeth with a toothpick, as well as rubbing them hard with a cloth: wherefore my teeth, back and front, remain as clean and white as falleth to the lot of few men of my years, and my gums (no matter how hard the salt be that I rub them with) never start bleeding. Yet notwithstanding, my teeth are not so cleaned thereby, but what there sticketh or groweth between some of my front ones and my grinders (whenever I inspected them with a magnifying mirror) a little white matter, which is as thick as if 'twere batter . . . I have . . . mixed it, at divers times, with clean rainwater (in which there were no animalcules), and also with spittle . . . and then I most always saw, with great wonder, that in the said matter there were many very little living animalcules, very prettily a-moving. The biggest sort had the shape of Fig. A: these had a very strong and swift motion, and shot through the ·. .ater (or spittle) like a pike does through water. These were most always few in number.

Dobell identifies this microbe as "a motile *Bacillus*." Taking into account size, shape, and the comparison with a pike that, as we saw, Leeuwenhoek had used before, I think Leeuwenhoek may again have seen a spirochete rotating so fast as it darted that he saw no sign of its coils; and he may easily have been so preoccupied with the complex world before his eyes as to miss what he had seen before: that when these microbes stopped for a moment they sometimes "bent their bodies serpent-wise." I have seen spirochetes behave this way.

To continue from the same letter:

The second sort had the shape of Fig. B. These oft-times spun round like a top, and every now and then took a course like that shown between C and D: and these were far more in number.

Dobell is certainly right in identifying this microbe as a spirillum. The teardrop shape is common, and its motility, described by modern observers as "tumbling," is very characteristic. What seems to be the same bacterium takes different shapes, with stronger curves that may turn into a spiral or helix, so that the microbe depicted by Leeuwenhoek in Figure G may be the same one. These longer microbes no longer tumble, nor are their curves obscured as they move. The shorter forms are often found in tooth scrapings in great profusion and wild activity.

Leeuwenhoek continues:

To the third sort I could assign no figure: for at times they seemed to be oblong, while anon they looked perfectly round. These were so small that I could see them no bigger than Fig. E: yet therewithal they went ahead so nimbly, and hovered so together, that you might imagine them to be a big swarm of gnats or flies, flying in and out among one another. These last seemed to me e'en as if there were, in my judgment, several thousand of 'em in an amount of water or spittle (mixed with the matter afore-said) no bigger than a sand-grain; albeit there were quite nine parts of water, or spittle, to one part of the matter that I took from betwixt my front teeth, or my grinders.

This description, taking into account that the smallest bacteria were barely visible to Leeuwenhoek, is compatible with a mixture of round cocci and rod-shaped bacilli in Brownian motion. More interesting to me is the estimate of concentration. Leeuwenhoek set the volume of his "sand grain" in an earlier letter (May 20, 1679) as a cube $\frac{1}{80}$ inch on edge. In metric units this would be 0.317 millimeter; and the volume, given by Leeuwenhoek as $\frac{1}{512.000}$ cubic inch, would be roughly 0.032 cubic millimeter. Since Leeuwenhoek estimates "several thousand of 'em," we will lose no precision by rounding the 0.032 to $\frac{1}{30}$ cubic millimeter. Taking into account his dilution factor (1 in 10), we can transform Leeuwenhoek's figure into "several" 10,000 $\times$ 30, or "several" 300,000—say, 1 million per cubic millimeter of original tooth scrapings. A recent estimate by modern counting methods yielded approximately 5 million cocci per milligram of scrapings from the gum cleft. Such a figure may vary by a factor of 10 either way, so that Leeuwenhoek may have been exactly right!

The same letter continues:

> Furthermore, the most part of this matter consisted of a huge number of little streaks, some greatly differing from others in their length, but of one and the same thickness withal; one being bent crooked, another straight, like Fig. F, and which lay dis-orderly ravelled together. And because I had formerly seen, in water, live animalcules that had the same form, I did make every endeavor to see if there was any life in them; but I could make out not the least motion, that looked like anything alive, in any of 'em.

This would have been a mixture of several filamentous bacteria typical of the mouth (*Leptotrichia, Actinomyces,* and others). If Leeuwen-hoek had thought of plants as well as animals, he might not have insisted that an organism was not alive because it did not move. This sort of bundle would be too big to show Brownian movement. The phrase "disorderly ravelled together" is beautifully descriptive.

A little farther along in the same letter we come to the most tanta-lizing of the figures, G:

While I was talking to an old man (who leads a sober life, and never drinks brandy or tobacco [Dobell says that "drinking" tobacco was the customary phrase then in Holland, and in England, too!], and very seldom any wine), my eye fell upon his teeth, which were all coated over; so I asked him when he had last cleaned his mouth? And I got for answer that he'd never washed his mouth in all his life. So I took some spittle out of his mouth and examined it; but I could find in it naught but what I had found in my own and other people's. I also took some of the matter that was lodged between and against his teeth, and mixing it with his own spit, and also with fair water (in which there were no animalcules), I found an unbelievably great company of living animalcules, a-swimming more nimbly than any I had ever seen up to this time. The biggest sort (whereof there were a great plenty) bent their body into curves in going forwards, as in Fig. G.

Dobell unhesitatingly identifies Figure G as a spirochete and indeed bases on it Leeuwenhoek's historical claim to the discovery of these microbes. I prefer to let that claim rest on the microorganism he described two years earlier in feces. In my opinion, which includes a full measure of respect for Leeuwenhoek's meticulous accuracy, G is too thick, the blunt end at the right is wrong, and the words "more nimbly," "biggest sort," and "bent their body into curves" don't fit together as a spirochete. I think this microbe was the longer variant of the tumbler shown in Figures B, C, and D—although "more nimbly" still doesn't quite fit. But spirochetes, even the thickest of them, may well have been a little too *thin* for Leeuwenhoek to have identified them in such detail with his equipment. It seems possible to me that he may have missed them in the mouth entirely.

Dobell mentions that several versions of this set of figures have come down to us and that the original has been lost. In the first Dutch edition of Leeuwenhoek's letters (1684), Figure G was missing; it is absent in the plate given by Schierbeek. The fact is that "matter lodged between and against" the teeth would be unlikely to show "an unbelievably great company" of spirochetes, the age and sobriety of the subject being immaterial. Leeuwenhoek evidently did not know how to go *down into the gum crevice* with his toothpick. It is down at the bottom of the pocket of pyorrhea, which is likely to be present in old people, that enormous accumulations of spirochetes are found. Poking into this region without a knowledge of its minute anatomy, which Leeuwenhoek did not have, is likely to elicit pain and blood;

and everything we know about the old master suggests that he would have avoided such maneuvers. That he did not try to bring up the sort of stuff that frightened the woman in the beginning of this chapter is shown by the paragraph just following the one quoted above, in which he described taking "matter that was lodged upon and betwixt the teeth" from another old man, this time one with teeth "uncommon foul" who testified that he never washed his mouth with water, "but it gets a good swill with wine or brandy every day." Reaching back through the centuries, I have no hesitation in asserting that this subject had plenty of spirochetes in his mouth; but Leeuwenhoek didn't find any.

Next comes a description of an experiment in which Leeuwenhoek found that "strong wine-vinegar" allowed to run between his own teeth, followed by rinsing "thrice with fair water," seemed to have little effect on the "unbelievable number of living animalcules" in material then examined under the microscope; whereas:

> I have also put a little wine-vinegar to this stuff mixed with spittle, or with water: whereupon the animalcules fell dead forthwith. And from this I drew the conclusion that the vinegar, when I filled my mouth with it, didn't penetrate through all the matter that is firmly lodged between the front teeth, or the grinders, and killed only those animalcules that were in the outermost parts of the white matter.

Thus the first account of antisepsis, two hundred years before Joseph Lister, and with a lesson that applies with full force today to every miracle mouthwash hawked on television.

But let us read on, in the same letter:

> I have had several gentlewomen in my house, who were keen on seeing the little eels in vinegar; but some of 'em were so disgusted at the spectacle, that they vowed they'd never use vinegar again. But what if one should tell such people in future, that there are more animals living in the scum on the teeth in a man's mouth, than there are men in a whole kingdom? especially in those who don't ever clean their teeth, whereby such a stench comes from the mouth of many of 'em, that you can scarce bear to talk to them: which is called by many people "having a stinking breath," though in sooth 'tis most always a stinking *mouth*.

This is followed by the quotation at the head of this chapter.

Leeuwenhoek was wise enough not to make my mistake of showing the mouth microbes to his curious but uninitiated neighbors. He himself was revolted by the stink from some of their mouths, but whether—again, as with disease—he did not think of associating the odor with the microbes, or for whatever reason, the microbes themselves bothered him not at all: he was clearly enthralled by them. In this general connection, Dobell refers to an early editor of Leeuwenhoek's works, S. Hoole, whose name appears in 1798 and again in 1807 and in later editions of the master's works. Dobell speaks of Hoole's "unusual squeamishness" and mentions, in connection with the brandy-swilling subject described above, that Hoole

> prudishly refrained from translating the observations made upon this disreputable "old gentleman," who . . . is called simply—without other descriptive detail—"a good fellow."

In listing Hoole's work in his bibliography Dobell says it was an

> excellent and scholarly translation (as far as it goes), but with all letters rearranged without reference to their originals, and with omission of all passages "which to many Readers might be offensive": consequently of little use to the student of protozoa and bacteria, and worthless to the student of spermatozoa—all reference to which is carefully castrated.

One more quotation from Leeuwenhoek, the end of the same letter of 1683:

> I noticed one of my back teeth, up against the gum, was coated with . . . matter for about the width of a horse-hair, where, to all appearance, it had not been scoured by the salt for a few days; and there were such an enormous number of living animalcules here, that I imagined I could see a good 1000 of 'em in a quantity of this material that was no bigger than a hundredth part of a sand-grain.

Leeuwenhoek's 0.01 sand-grain, using the same equivalent as before, would be roughly 0.00032 cubic millimeter. His "imagined" concentration of microbes—which sounds like the old master's version of a guess—would therefore be something like 3000 times his "good 1000," or 3 million microbes in a cubic millimeter. A cubic centimeter (or a milliliter) would therefore have contained some 3×10^9, or 3 billion, microbes. Accurate counts of the bacteria in the soft white deposit on teeth have not been made, but I would expect them to be similar to or higher than those for saliva, in which counts of this magnitude are common enough. The variation may be tenfold or more either way, as before, so that Leeuwenhoek's guess is within the scope of truth to a degree that takes my breath away.

3

The Facts of Life
on the Earth

The oyster being achieved, the next thing to be arranged for in the preparation of the world for man was fish. Fish and coal—to fry it with.

—Mark Twain, *The Damned Human Race*

The appearance of life on a place like the earth is thought to be inevitable, given appropriate conditions and enough time. Even with appropriate conditions, a long succession of accidents is called for; the combination is spoken of as the sort of highly improbable event that happens if one waits long enough. We think life is inevitable in outer space as well as on earth because, astronomers tell us, among the countless planets in space some must have life-permitting conditions like those of Earth. Time is available everywhere. We are coming to have a pretty good idea of how life started on Earth, and the idea gives us courage to speculate on the existence of life elsewhere.

There is a rough parallel between the origin of life on Earth and its origin on man. There is comparable inevitability, but with the probability of its development shifting from remote to immediate, that life will appear on any part of man—his mouth, for instance— given, again, appropriate conditions and time. The conditions come down from billions of years ago to now, from a developing earth's surface to the biological terrain of a new baby; and the time scale changes from vast stretches of years measured with exponents to days, hours, minutes. One prospect is remote and wonderful, the other intimate and creepy. They are parts of the same Scheme.

23

Within my memory, ideas about the origin of life have changed and grown. Bacteriologists were conditioned by Pasteur's immensely fruitful rule that life could not originate from anything but pre-existing life, and tended to be blocked on the matter of origin. Under the pressing circumstances of a burgeoning science, which bacteriology was up to about thirty years ago, it hardly occurred to its founders or to my colleagues and myself when I was younger that the rule needed modifying: life springs only from life *under conditions now prevailing.*

That mice could come out of nothing but corn and a dirty shirt in a box in the dark, maggots could be produced in decaying meat, cockroaches appear without the efforts of parents in food scraps undisturbed in the kitchen—things like this were once believed by scientists as well as by presumably more ignorant folk. Even before Pasteur's time experiments had forced most of those who knew about them to think of the spontaneously generated animals as smaller and smaller. Pasteur demonstrated that even microbes did not originate without pre-existing microbes and so scotched the whole notion. He made it clear that microbes are full members of the living world, distinct species, breeding true in the same way as corn or cats; and by doing this he opened up a vast new area of biology. The American chemist Melvin Calvin, a noted contributor to our knowledge of the origin of life, reminds us that when Pasteur did this, however, he also put an end for a time to the consequences of certain ideas expressed by his great contemporary, the other giant of biology, Charles Darwin. Calvin quotes the following from a letter of Darwin's dated 1871:

> It is often said that all the conditions for the first production of a living organism are now present, which could ever have been present. But if (and oh! what a big if!) we could conceive in some warm little pond, with all sorts of ammonia and phosphate salts, light, heat, electricity, etc., present, that a proteine compound was chemically formed ready to undergo still more complex changes, at the present day such matter would be instantly devoured or absorbed, which would not have been the case before living creatures were formed.

By the 1930s bacteriologists had come to recognize that bacteria and viruses, the things they spent most of their time with, were probably the smallest things to which the notion of life could be applied. It began to appear probable that something of that sort might have been the primordial living thing, the first fully formed living thing to appear on earth. Since 1935, when a virus was crystallized for the first time by Wendell Stanley, and especially in recent years, it has become plain that all viruses are chemically and structurally different from bacteria and from all living cells. Their intimate relationship with cells—apart from which they cannot multiply—is such that they could not have been present on earth *before* the cells on which they depend or of which they form-parts. It is possible that the first virus, or group of viruses, was present in the first cell. If not, viruses must have evolved later.

This line of thought, developing slowly, tended to leave the bacteria—assuming *their* earliest ancestors to have been bacteria too—as the group within which the primordial living thing was to be found. Speculation tended to fix on bacteria but was inhibited and even distorted by a narrow construction of Pasteur's rule. At about that time T. S. Eliot was suggesting that the world would end *"Not with a bang but a whimper"*; yet it hardly occurred to us that life might have begun in just such a way. It seems strange now to realize that so many of us could have accepted the notion that life might have originated like Minerva, who sprang fully formed and armed from Jove's brow; that where there had been no life before, lo! it existed. It is hard to know whether even Darwin, the father of the theory of evolution, fully realized how *gradual* the process might—we think now, *must*—have been. From his words just quoted, it is possible to imagine that, once a protein was formed, the first cell completed itself like a closing zipper.

But Darwin had the essence of the modern view when he recognized that organic material could accumulate "before living creatures were formed." Our difficulty in the 1930s centered on this point. If life had emerged soon after the first spontaneous synthesis of organic compounds, the new life would have had little or no organic matter to feed on, and it seemed to follow that the first living thing must therefore have been able to manufacture all its own substance, given nothing but elements and inorganic materials to work with. Bacteria with this sort of capacity do in fact exist today. They get the energy they need either much as green plants do, from the sun by photosynthesis, or by being able to bring about energy-producing changes (oxidations) in inorganic substances—hydrogen gas, reduced iron salts, ammonia, sulfur, or hydrogen sulfide. Such bacteria have the complex machinery needed to fabricate the whole range of substances found in cells—proteins, carbohydrates, fatlike substances, nucleic acids, and vitaminlike compounds—which, with water and other simple materials, make up the cell. The idea that the first cell could have been so clever seems by hindsight not much of an improvement over the Bible story.

A more reasonable approach to the origin of life was being thought out during this period, starting with the Russian biochemist A. I. Oparin and the British biologist J. B. S. Haldane. Oparin proposed first, in 1924, a scheme of *chemical evolution* preceding and making possible the first appearance on earth of an organized, self-sustaining, and self-reproducing body that could unhesitatingly be called alive. During this chemical period, since there was as yet no life, natural "spontaneous" organic synthesis could take place, and increasingly complex substances could be formed and might accumulate. Today, as Darwin knew, any naturally fabricated organic compound in the free environment is rapidly found by a living thing that can make use of it, and is accordingly destroyed for its food energy or incorporated into the substance of the living thing. In either case the compound is removed from the environment.

Oparin argued that when the sort of complex self-energizing reproducing body we call alive first appeared on earth, it could have been not the excessively complicated cell that could build its substance out of elements and simple salts, but a more deficient, slower, and therefore simpler thing that needed organic compounds already present and available as its food.

At first this idea had rough going. The miracle of a bacterial Minerva was more acceptable, if only because the whole subject was only lightly thought of, being, presumably, un-Pasteurian. Even twenty years later Oparin's hypothesis got little more than passing mention— in one case, in a footnote, as a possible alternative to the miracle. But a few people saw it as a more likely idea than that of a sudden leap out of the void. In 1953 Stanley L. Miller, then a graduate student working with the geochemist Harold C. Urey at the University of Chicago, brought out experimental evidence for Oparin's idea and, almost overnight, changed the whole drift of thought in this field. Miller was able to make significant organic compounds—amino acids, the structural units of proteins—by setting off an electric spark in a sealed vessel containing only substances believed to have been present in the atmosphere of the young earth: methane, hydrogen, ammonia, and water. The spark lent the energy for the synthesis and represented a primeval thunderstorm.

Since that time much more experimentation has fortified the idea of a long period of chemical evolution preceding and merging with a period of biological evolution. The first may have lasted as long as a billion (10^9) years, and Calvin puts the second at another two billion years. In the beginning the earth's atmosphere probably contained little carbon dioxide and no free oxygen gas. Carbon dioxide is needed for the photosynthetic activity of green plants, and oxygen is necessary for higher animals to breathe. Plants and animals would have come into existence in that order. The first living things got along without oxygen, as many bacteria and other microbes still do. Pasteur, who discovered this phenomenon, called it "life without air." Instead of oxygen, these microbes use means of burning food for energy which are much slower and more wasteful but which serve, and carbon dioxide is often made in the process. This carbon dioxide, with water and salts and sunlight, is all that green plants need for their photosynthesis, from which oxygen comes as a by-product. The energy released in fire and in the muscular movements of animals needs oxygen gas.

Spontaneous generation of life, as Pasteur showed, cannot occur on earth today—or, if by remote chance it could, it cannot lead to anything productive. The universal presence of life on earth prevents the accumulation of organic material. Living things find such material and decompose it. Animals break down plants and in their turn are taken apart by other animals. Microbes convert the carcasses of both plants and animals back to a point at which plants can start over again. But spontaneous generation could have happened 2 billion years ago, given a further billion, more or less, for the needed organic material to have formed and piled up. In the beginning and in due order of succession thereafter, with plenty of time for slips and failures and fresh starts, the necessary materials came to be present, the permitting temperatures arrived, and energy to make the process go burst forth. The energy, as you must know, came ultimately from the sun, more immediately from lightning, and gradually from chemical sources as a supply got stored in the compounds that were accumulating. A billion years or so of such organic chemical accumulation were required before Darwin's "still more complex changes"—which must have happened countless times before they succeeded—finally led, once or many times, to the energy-producing and self-perpetuating thing we call life. And at some point the foothold of life was secured and it began to climb the scale of organization. After 2 billion years or thereabouts, man appeared. If we think of the whole process in a scale reduced to one year, we have been around for about an hour.

This sort of scheme of the origin of life and of human beings has come to be generally accepted only during the last decade, and, as I have said, the concept didn't have easy going. It bears on our wider purpose to speculate a little more as to the reasons for opposition to it. Unlike the strenuous opposition to Darwinism a century earlier, the trouble this time arose from scientists themselves, rather than from anyone who, in the view down the scientist's nose, could be called bigoted. Part of the difficulty may have had to do with Oparin's being Russian and with Haldane's British leftism. But a preference for a miraculous event at a time when all doubts of the general validity of evolution had disappeared among biologists (except perhaps in Tennessee and one or two other places) must have had something more than political roots. Prejudice was at work again.

One of the common consequences of prejudice is irrational discrimination—a preference for certain members of a class for reasons that will not withstand dispassionate scrutiny. Perhaps all such notions can be traced back to matters of personal vanity, or to man's common need to reassure himself of his own worth. How we come to question that worth in the first place is a recurring phase of the fundamental problem.

The particular prejudice I am now speaking of may have come, through such self-doubt, to a matter of discrimination by bacteriologists among bacteria. This is nonsense, of course; but so is all prejudice, I suggest, when one looks at it in a good light. In this case it may be that we boggled at the notion that a particular class of bacteria, which had attributes with disagreeable overtones, could represent our own remote ancestor, the primordial cell. Although I am sure nobody thought in these terms, it is possible to imagine the denial of even the unformed idea of something much worse than having a murderer in one's family tree. Let us examine the point further.

Oparin rejected the idea that the prototype cell resembled today's free-living bacterium—called "autotrophic," (self-feeding) or sometimes "lithotrophic," (feeding on minerals). Such a cell synthesizes all its substance from simple materials. He selected instead a kind of bacterium that needs organic foods and has correspondingly limited powers of synthesis—"heterotrophic" or "organotrophic" (feeding on others or on organic materials). Of course he did not suggest that the primordium could have been just like any bacterium known today. But both lithotrophic and organotrophic bacteria do exist, and it is natural to use what we know as models for a projection of what might have been.

But today's organotrophic bacteria, especially those that are structurally simplest and have the least capacity to make cell substance, have a property that Oparin's choice could not have shared. In a world without other living things, the primordium was necessarily free-living although organotrophic. The combination was made possible by the accumulation of organic material synthesized during the period of chemical evolution. But as life developed it used up this chemical reserve, and later the organic supply was replenished only through the work of living things themselves. Today's bacteria of limited synthetic power are all dependent on other life, and at their most dependent, when they become structurally and functionally simple enough to fit Oparin's suggested specifications, those we know today are all *parasites*.

If evolution began with microbes of some such sort, they would necessarily have differed from parasites in lacking those attributes of parasitism that enable such organisms to survive within their hosts, since at the time there were no hosts. Hence we can be sure that the primordium could not have persisted in its original form, but that it must have disappeared as evolution proceeded, as many organisms are known to have evolved and disappeared in the intervening centuries. Today's parasites must have evolved toward parasitism as their plant and animal hosts evolved, by a process thought of as regressive— in which one function after another was lost, together with its chemical machinery, as the host supplied its purpose or furnished its product ready-made. Since the host seldom accepted this arrangement passively, the parasite also evolved means to protect itself against the host's objections. But on the whole the evolution of parasites is a downhill process, and the word "degeneration" seems appropriate for it. The unavoidable opprobrium in this word is associated in our minds with a distaste for blood-sucking vampires, or for the "creeping, venomed thing" that Shakespeare's Anne called Richard Gloucester before she married him. The notion and its overtones are also associated, among bacteriologists like everybody else, with the length and breadth of infection and disease.

In the 1930s, considering that there were more important things for them to think about, bacteriologists tended to dismiss out of hand the idea that we could have had such an ancestor. What the monkey was to Bishop Wilberforce and William Jennings Bryan, the parasite may have been to my colleagues. If this be true, the point becomes the more droll because in botn instances the very image was distorted: it couldn't have been a monkey, and it couldn't have been a parasite.

If this was in fact the kind of prejudice that slowed our approach to the earliest phase of evolution, it has by now disappeared among scientists. The path is therefore clear to let us dispose of one prejudice and to provide us with a realistic answer to the child's most basic question: "Where did I come from?" And the truth, if not stranger, is certainly more wonderful than the myth.

We have not descended from a parasite, but it would be no shame if we had. All life is a single community, including the parasites. Some of them are our natural enemies and need to be treated accordingly. But it would be good for us in general—and would help us get to the next chapter—if we could stop thinking indiscriminately of microbes and parasites as repulsive, contemptible, or ferocious. These adjectives apply with more force to later products of evolution and doubtless fit most snugly on man himself, or anyway on some of his behavior. Even with regard to man—or, if you like, mainly with regard to man— we need to define what is repulsive and distinguish it from what is not. Our own parasites, especially those that are always with us and part of us, call for the same discriminating appraisal. We will not stop trying to exterminate some of them and we will always be interested in controlling them all as far as we can; but there are better reasons for doing these things than those that spring through prejudice from ignorance.

4

This Is the Life

... perhaps the effort made in this book ... will serve another end besides the more obvious ones—as a contribution, however small, to human emancipation.

—Theodor Rosebury, *Microorganisms Indigenous to Man*

As life seems to have been inevitable on earth, so it seems to be inevitable on you and me, the conditions being appropriate in both instances, and the time being available. But life doesn't evolve on each person as it has evidently done on earth. It finds its way to us from outside, much as early man is thought to have come to America across a land bridge in what is now the Bering Sea and migrated to habitable regions and colonized them.

The life on man consists of microbes, microbes in extraordinary variety and in large numbers. I once counted some 80 distinguishable kinds in the mouth, and the total number of bacteria excreted in feces by an adult each day ranges under normal conditions from 10^{11} to 10^{14}—from 100 billion to 100 trillion. The microbes of our normal population inhabit nearly every surface that is freely exposed, such as the skin, or accessible from the outside, such as the lining of the intestinal tract. Along the length of the alimentary canal, from mouth to anus, some of them grow in particles of food or in what remains of food as it undergoes digestion. These microbes, speaking very strictly, are not parasites, or may not be; they merely take advantage of the warmth and moisture of the environment to live on our food or the products we digest it to, growing as they might grow on similar materials in an incubator. Some bacteria grow on the products produced by the activities of other bacteria. But more characteristically the life on man, including many of the microbes I have just spoken of, is made up of parasites, which grow in more intimate relation to our tissues. Many of these microbes are unable to grow elsewhere in nature. Most of them can grow *within* the body as well as on its surface, when conditions permit. Such conditions imply that something is wrong with the host. The result is disease, a special sort of disease in which the ordinarily harmless, sometimes even beneficial surface parasites take part. We cannot say much about the life on man without recognizing that a capacity to damage him is latent in it. The poorly defined state we call health consists in part of a balance maintained between ourselves and our microbic populations. This point and another need to be mentioned to avoid the appearance later of being contradictory. The other point is that the number of microbes that make up the life on man, although large in aggregate and in certain places—such as feces and parts of the mouth—is small compared with what it may become under abnormal conditions. We may defer these matters until a later chapter. Our subject now is life on man in health.

Microbes are a pervasive part of the living world. They are present on land and in water, in soil, on healthy plants and animals, and in the carcasses of both. It is the microbes, as you must know, that do the work of returning plant and animal remains into the cycles of nature, breaking down the dead tissues so that plants can use their components again. Microbes are also found in the air, but mainly as they are spewed into it by the activities of animals, especially man. It was Pasteur who showed us that as we go up mountains into higher air the bacteria get fewer and fewer. The ultraviolet rays in sunlight tend to kill them. Above the ozone layer of the atmosphere, high above the clouds, the concentration of short-wave radiation is much greater, and microbial life is probably entirely absent.

All healthy animals and plants have microbes growing on surfaces exposed to the environment directly, such as the skin of a grape, or those exposed indirectly, such as the lining of your intestinal tract. But under these surfaces, within the tissues of plant or animal—unless something is amiss there—microbes penetrate only to be destroyed quickly; otherwise infectious disease results. Yet microbes don't remain entirely outside. Skin and the intestinal wall have a certain depth, including folds and natural (microscopic) pockets and recesses, all lined with cells called epithelium, into which the microbes can burrow and where they grow while the host stays healthy. In one way or another a few are continually delving deeper and persisting under the epithelium for various brief intervals—hardly more than an hour or so—while various bodily police and scavenging services sweep them up and destroy them. There is a sort of absolute ban on immigration inside a barrier zone of integument. The barrier is active all the time, like a very slow filter in constant use. Normally anything foreign that gets through and has a size and structure typical of the complex molecules of living cells—proteins, for instance—runs into biological machinery especially adapted to recognize that it is foreign and to get rid of it. This machinery supplements that of digestion, whereby the big molecules of food are broken down to small ones that are acceptable to the tissues. Both mechanisms are highly effective but not perfect; but breakdown is abnormal, and, again, we must not let ourselves be tempted to speak of it further.

The fetus is part of its mother's body and has no microbes in its sheltered state. When the time comes, it breaks through into contact with the world outside, shedding its aquatic habit together with amnion and placenta, and becoming terrestrial, much as a tadpole does when it grows into a frog. Until this time it is protected from the microbic world as the mother's own tissues are protected under and by her surface barriers. But as the normal fetus changes into a baby and pushes its way through the stretched vagina it begins at once to pick up microbes that grow there; and with the first crying gasp that follows the obstetrician's slap, be the environment clean as aseptic technique can make it, it draws microbes into its little nose and mouth and throat.

But the microbes arrive slowly, and normally this first contingent is small. The microbic population that will come to inhabit the new human arrival on earth is by no means simply a hand-me-down from mother to babe, and the notion that the infant is blanketed with microbes on contact with infested air is entirely wide of the mark. Microbes there are, certainly; but the mother's birth tract is normally only lightly settled, and the initial environment of hands and air and blankets is likely to range from clean to nearly, if never quite, sterile.

Actively disease-producing bacteria can cause trouble if they happen to be around, and one of the important jobs of modern sanitation is to keep them away. If the bacteria that land on the baby are not of this active disease-producing (pathogenic) kind—and most bacteria are not—many of those that do land on his skin or get drawn into his mouth or nose are likely to die there. This process is a detail of the vast pattern woven by the trial and error of evolution, with an assist here from human intervention. The advent of microbes is one of the events that make birth a crucial point in life, and meticulously selected genes and boiling water have advanced the cause of survival.

We have every reason to believe that the microbes arrive in an orderly succession, depending on progressively increasing contact of the baby with its environment and on the baby's continual acquisition of habits, structures, and mechanisms that work both ways: some encourage immigration, some ward it off. We don't always know just where the microbes come from. We don't see them arrive, and they don't register their point of origin as immigrants to this country used to do at Ellis Island in New York Harbor. At certain times after birth and in certain places we find certain microbes when we look for them. We don't always find the same ones, but what we find is similar enough from one person to another so that something like a design becomes plain.

In the breast-fed baby one of the remarkable first arrivals is a bacterium that looks like a short forked stick and is appropriately called by the specific name *bifidus*. During the nursing period it makes up from 90 to 99 per cent or more of the infant's fecal population. Breast-fed babies continue to fare better, by and large, than comparable infants on cow-milk formulas, and bottle-fed babies have a mixed microbic population in their intestinal tracts. The bifid microbes have accordingly been given credit for salutary effects; but attempts to define the relationships, if only in the hope of making up a formula as good as mothers' milk, have got nowhere. The argument that something in mothers' milk sustains *bifidus*, which in turn keeps more harmful bacteria at bay, broke down when it appeared that the forked sticks are just as numerous in bottle-fed infants as in nurslings, but bottle babies have enormous numbers of other species too. Nor is it clear that these other bacteria are in fact harmful. We can be reasonably sure that *bifidus* is harmless, but that's about all.

The puzzle is further complicated by our uncertainty as to where the bifid bacteria come from in the first place. They don't seem to be widespread in nature outside of the nursling stool. They are found in the vagina of pregnant women, in the pre-nursing breast fluid (colostrum), and on the nipple skin of lactating mothers. Caesarian nurslings have them, just as normally born infants do, so the source is probably the nipple, but how do they get there? They are not otherwise known to be native to human skin. No doubt these questions will find answers in time.

But while we remain ignorant of details, we have only to watch our infant sons and daughters, with the balanced caution and tolerance that comes of love and understanding, to have a pretty good idea of the sources of their normal microbes. And such specific information as research has provided tends to support our inferences. The healthy baby finds the external world fascinating. He explores what we approve of as well as what we think of as repulsive. Every new taste, smell, sound, sight, and feeling, limited only by pain and parental restraint, is a new source of excitement and of both education and—given only contact—microbes. During what Freud called the "oral" phase of development, as the pleasure of suckling is tempered with growing awareness that there are other pleasant objects as well as the nipple, the world is sampled largely through the mouth, and the world of microbes enters there and sorts itself out along the alimentary tract. We infer from what we find in our cultures that most of the bacteria of the external world—those of garden soil, for instance, which certainly find their way in—meet an inhospitable environment and are more than likely to be dead when they reach the anus. We recover them only occasionally in cultures of the stool, or their numbers are insignificant compared with the more typical stool species. There is no great puzzle here; the species that find the habitat suitable are most likely to have come from a similar one; they settle down and proliferate. Others may merely survive the ride, or they may not.

Yet the important conclusion we feel justified in drawing from our analysis of microbic species in man is that since so many of them appear in man and not otherwise in nature, they must be transferred more or less directly from man to man. Hence if, as I say, the baby does not simply acquire its microbic population from its mother during the process of birth—and we are sure it doesn't—it must get it later from other people, mother doubtless among them. The rule is not absolute or invariable. What is true of normal microbes is also true of abnormal ones—the disease-producers, or pathogens that usually have no permanent home in healthy people. A process of adaptation to man is characteristic of microbic parasites as of larger ones. The adaptive process depends partly on the presence in the host environment of food the parasite requires (and other sustaining conditions like temperature and degree of acidity), as well as on the ability of the parasite to fend off the protective or parasite-destroying devices of the host. Adaptation proceeds actively on both sides during the course of colonization: of parasite to host and of host to parasite. Part of the host's adaptive process consists in the development of antibodies to those big microbic molecules that get through the integumentary barriers every now and then. Having destroyed them the first time, perhaps laboriously, the host has learned how to do it faster and more efficiently when it happens again. Thus a relative immunity is achieved.

All these processes have a kind of relativity that it is unwise to oversimplify. Between parasites and what we call saprophytes—microbes that grow in the food mass only, not in the tissues—there is no sharp line. Not only the parasites but, curious though it be, the saprophytes themselves, blend without a sharp boundary with the pathogens. Generally speaking, the normal microbes do not damage the host if he is not already damaged in other ways, it being understood that interaction between the two never ceases while both remain alive. Continual barrier penetration and, in consequence, accumulating tissue experience with large-molecule microbic components and its associated immunity—directed with extraordinary refinement against the particular molecules of each microbic variety separately—keep the microbes at the barrier, doubtless limit their population size, and protect the host tissues, all relatively, in degree varying with microbic species and host condition. The process is not unfamiliar. We ourselves, macroorganisms that we are, become adapted to the macroenvironment in which we live at the cost of never-ending interaction, which continually modifies us and it. Such things vary only in detail across the span of the living world. Life proceeds thus at all its levels. The process can be pleasant as well as unpleasant, and we have no right to assume that a boundary to pleasure can be found with the microscope.

Born without microbes, we acquire a normal microbic population in a way whose relative consistency is all the more remarkable in view of all the possibilities there are for variation. Microbes unlimited visit us from everywhere in the air we breathe and in the food and water we take in, but mainly by contact, and most significantly by contact with other human beings. Far and away the greatest proportion of visitors goes in through our mouths; and the presence there and farther along in the rest of the digestive tube of the food we eat and the products our digestive enzymes make out of it assures the richest microbic populations we harbor. Even so, and varied as these populations become after the period of mothers' milk, they are not infinitely varied. They are a characteristically selected group, characteristic in their kinds and in their relative numbers.

It has been thought for a long time that the alimentary tract is most densely populated near its openings in the mouth and in the large intestine from its beginning in the cecum on down to the anus. The stomach has a strongly acid secretion, and the finding of few microbes in the juice of the empty stomach is usually interpreted to mean that it is an inhospitable growing place for microbes. Samples of the scanty fluid in the healthy empty small intestine, taken with syringes during surgery in adjacent areas, have also yielded either no microbes or a scattering of mouth species. It has been suggested that the active downward-moving contractions of this tube that we call peristalsis help to keep it microbe-free when the heavily populated food bolus has passed by. But more recent experiments with mice show the *walls* of the tube rather than the fluid in it to be teeming with microbes. The parallel experiment is, of course, harder to do with human subjects than with mice. The single attempt I know of— to study the bacteria within the wall as well as in the juice of the upper region of the intestine—turned out to support the earlier finding of a sparse microbic population and so suggests that man and mouse are different in this respect. More work will be needed to settle the question. Meanwhile the idea of microbes living like troglodytes *within* the walls of man rather than just *upon* them certainly remains valid in other areas. Most of what we know of this subject is superficial in the strict sense; it is easier to study the surface than the depths of living tissue (as, no doubt, of most other things). But we know that we can scrape the tongue and keep on recovering bacteria in seemingly inexhaustible supply. We can scrub the skin with soap and water until pain makes us stop, and the microbes keep coming. Whatever may be true of the upper end of the alimentary canal, in its lower reaches the true microbic population is likely to be as different from the population of voided feces as it is harder to get at.

The digestive system is not the only place where the normal microbic population lives. Nor is the food we eat the only nourishment we provide them. As I mentioned before, there are microbes capable of using as food for themselves any naturally occurring organic substance. All our surfaces are continually discarding such substances, in surface scales and cells, in mucus and other secretions and excretions. Apart from the food we ourselves eat and its digestive products, these materials are sufficiently similar from one moist inner (mucous membrane) surface to another so that a basic population is found in all of them, with local variations. Skin differs from the moist surfaces in its relative dryness and in the fatty substances of its special secretions, and its microbes are correspondingly distinctive. No body surface lacks microbic nutrients, whether the nutrients come from without or within or both.

That certain regions easily accessible to microbes are nevertheless only sparsely inhabited results from particular local conditions that actively oppose habitation rather than from food scarcity, with the exception that microbic concentrations tend to be much lower generally where external food is absent. But in such areas—apart from the alimentary tract, that is—the variation has little to do with our prejudices: neither the microbes themselves, nor the conditions for or against their colonization, which are set dispassionately by the structure and function of our bodies, have any regard for what we think of as modesty.

The moist exposed surface of the eye, for instance, is kept nearly free from microbes by a combination of winking, washing with tears, and antimicrobial agents in the tears, these things tending to destroy microbes and to wash them into the nasal passages. The respiratory tract has protective devices that keep its microbic population down. From a point just inside the hairs in the nostrils back to the upper pharynx the nasal passages have a microbic population, but not a lush one. Sneezing is one of several nose-clearing devices. The throat, especially that part of it that leads to the esophagus and forms a common corridor with the alimentary tract, is more densely populated, with the recesses in the tonsils, which lie just behind the mouth, harboring a group of microbes much like that of the gums. Farther down, in the larynx and the inverted tree of diminishing bronchial tubes, the population thins out rapidly until complete sterility is achieved; this is maintained by a complex battery of housekeeping devices, including minute hairlike projections on all the surface cells (cilia), continually brushing upward, the cough reflex, and a number of cellular activities and antagonistic materials.

The skin is distinctively populated, with interesting variations in different places. The hairless areas of face and hands, which are most exposed to the world, tend to have a scantier normal population than covered skin has, greater exposure to external contamination being offset perhaps by drying and other effects of exposure. This does not apply, as you would suspect, to the environment-catching portions of the fingernails, in which, for example in children, one expects and finds a special set of microbes often more like those of garden soil than anything else. Sweaty or oily places such as the angles of the nose, the armpits, the creases under women's breasts, and the spaces between the toes, have a richer population than others. The opening to the ear is the habitat of a goodly mixture of microbes. I know of only a single report of bacteriological studies that included the navel, in which eleven men were examined under conditions simulating those in a spaceship, to see how they might accumulate microbes. The navel was not studied intensively, and the findings are not remarkable. The most significant accumulations of microbes were found in the armpit, the groin, and, curiously enough, on the penis. I know of no data on men under more normal circumstances with which to compare these findings, but they are sure to have been influenced by the conditions of confinement and special clothing.

What we know of the genitourinary tract normally, from the external cloacal area to the tubes and structures within, is similar to the pattern in the respiratory tract. The sweaty folds between the scrotum and the thighs, and generally the groin region in both sexes, resemble the area of the armpit except for continual contamination from feces; but the characteristically fecal microbes do not flourish there. The mixture of distintegrating skin cells and secretions called smegma, found under the prepuce and around the clitoris, is well populated. The vulva otherwise is like scrotal skin in being exposed to continual fecal contamination without marked effect on its native population. Its microbes are more similar to those of the tonsillar area, but they tend to be fewer. Farther inside, in both sexes, the microbes fall off rapidly. There is a sparse collection at the outer end of the urethra, quickly diminishing to sterility toward the bladder, which, with its contained urine, is germ-free in the healthy state. Normally urine becomes contaminated as it is voided, but only lightly. That is, fresh healthy urine contains few microbes; but, as our noses inform us, these few multiply rapidly in standing urine, which Pasteur found to be an excellent culture medium. Contamination of saliva and sweat takes place similarly. Both are free from microbes as they emerge from their secreting glands. Sweat changes as urine does, the resulting products and the smells being different. With urine, a dominant product is ammonia, produced from urea. On skin the fatty components of sebum contribute the rancid smell of fatty acids. Saliva helps to wash microbes from mouth to stomach; spat out of the mouth, it is a rich microbic suspension at once, containing under healthy conditions some 10^8 to 10^9 bacteria per milliliter.

The healthy vagina harbors a sparse collection of microbes which undergoes characteristic changes during puberty and menopause. Beyond it, the uterus, like the bladder and the lungs, is—not sterile this time—free from microbes. In the newborn infant hormonal effects similar to those of the mother encourage vaginal colonization of a restricted range of microbes like those of the maternal vagina. After the first month of life and up to puberty, the vaginal secretion is slightly alkaline, and a diverse microbic population is found there, normally not great in aggregate. The appearance in the vaginal wall at puberty, evidently under hormonal influence, of animal starch (glycogen) is associated in some indirect way with a sudden reversion to a simple group of microbes which are typical fermenters. As a result of their action the vaginal secretion becomes acid. This acidity is credited with preventing the growth of a more mixed or abundant microbic population, and hence with preventing infection. What happens is probably not so simple as we once thought, but a protective mechanism ultimately under the control of the female sex hormones evidently operates in the healthy vagina during the period when sexual activity, and in particular childbearing, would make it most needed. Certain acid-producing microbes evidently contribute their beneficent influence. The effects are exerted against a range of common infections not including the venereal diseases, suggesting that venereal diseases may have come along relatively late in the evolution of this phenomenon, or after it had been established. After the menopause, conditions return to those of the presexual period: a scanty alkaline secretion and a sparse, varied microbic population.

I have said that microbes come to us mainly from other people, and have suggested in support of this idea, first, that we are born without microbes, and, second, that the kinds of microbes we have are more like those of other people than they are like microbes from any other source. I also mentioned in passing that there are similarities among mammals, and it is possible that some of our microbes come to us from animals, and vice versa. There is additional testimony bearing on the idea that our microbes come from our neighbors as part of our cultural inheritance. The evidence is somewhat indirect, especially in that it deals not with normal microbes but with pathogens, and tells us more of the means of transit of the microbes than of the microbes themselves. Necessity has dictated closer examination of pathogens than of normal microbes. Studies of their transmission pathways are obviously part of the exploration of epidemics. Refined methods have been developed, based on the precise marking of microbes (and of viruses), comparable to fingerprinting and blood-typing for the identification of individual men. But with microbes it is not individual germs but subspecies varieties—"types" is the technical word—that are identified.

By such methods we have been able to trace, for instance, the three types of polio virus through a complete cycle, including details of the course of infection in a child and the transfer of virus in the external world from intestine to anus to another child's mouth. The sequence, anus-to-finger-to-mouth, has proved to be commoner than "nice people" like to believe. The contamination of bathing places from all the body's orifices is another commonplace to which polite people blind themselves to their cost.

Among microbes proper we know from many studies of types of staphylococci and certain wild or abnormal colon bacilli, among other bacteria that cause disease among infants and others in hospitals, that their source is other people. Hospital staff people are the usual transmitters, especially those that attend the patient most closely. Transfer from mother to baby happens more often with colon bacilli than with staphylococci. In fact with these latter the infected infant is more likely to infect the mother while suckling than to be infected by her. But these details concern us less than the fact that the routes of transmission are available and open, and that they evidently remain open in spite of elaborate attempts to close them in today's hospitals, where such "nosocomial" disease is a continuing problem. The root of the problem is the accident that pathogens have come to usurp space in *part* of the population ordinarily reserved for more tolerable parasites, and have got somewhat out of control by entering the transmission mechanism of the normal microbes, causing trouble as they land on territory unfamiliar with them. The transmission mechanism is not a problem with the normal microbes.

If we can protect the infant (and older persons, too) against the pathogens, he will take care of the normal microbes himself. As he acquires them group by group, as they enter and migrate and establish themselves and die off and are replaced—a process that starts at birth and goes on through life—he grows accustomed to them. He never stops reacting to their presence, and in so doing he learns to keep them in their places—to live with them. They never lose some degree of danger to him. On the other hand some of them are clearly helpful, as we have seen; and more details as to both the help they give and the harm they do are to come in later chapters. But it begins to appear that man is better off, on balance, for the presence of his microbes, just as he is better off for being part of a world of living things larger than microbes.

5

At Home on Man

Life would not long remain possible in the absence of microbes.

—Louis Pasteur

If we could assume a perspective like Gulliver's in Brobdingnag, and approach a person—any reasonably healthy person—as we hope someday to approach a cousin planet circling a brother of our sun, and if we could land on him and survive, we might be entranced with what we found, as he was. But the living things we would find do not resemble man in the least, and they are so small that Swift's twelvefold reduction wouldn't be nearly enough to let us see them. Make it a thousandfold and we might see bacteria as they appear under a microscope that magnifies that much. Picture yourself if you can as about one-fourteenth of an inch (less than 2 millimeters) tall, which is as tall as this letter I. The average bacillus would look to you like a little stick, less than a millimeter thick and about five millimeters long.

But the microscope distorts the dimension of depth and almost abolishes natural color. If we could see microbes directly they would have both. I think their translucency, and colors both from their pigments and from prismatic effects, might make of what we saw as fascinating a new world as the one Leeuwenhoek found. If we felt quite safe there, we might react with wonder and delight.

Let us relieve our object person of his clothes—he is naked!—and land on a gently curving plain of relatively hairless skin, say a shoulder blade. We could hardly miss the growing things, but our first impression, encouraging the feeling of safety and hence the favorable reaction, would be one of plants rather than animals. We would probably find no creeping, crawling, or darting things here. The ground, made up of flat stones like slate but whitish, would show layers of more translucent stones beneath, and we might get a glimpse of depths of color beneath—pink, brownish, or purplish, depending on whether we had landed on light or dark skin. Among the stones would be standing hairs and the open pits of sweat and sebaceous glands. Grouped around the bases of the hairs and at the mouths of the sebaceous glands we would see clusters of beads, still tiny even to our diminished vision, like very fine caviar, but mainly whitish or opalescent, some with a yellowish tinge or clearly amber-like, all probably shimmering with iridescent color as well. We would see bundles of soft translucent sticks, stalks, and twiglike things. There would probably be no animal-like movement at all, but something more than the passive movement of plants swaying in the breeze on earth would be there. We would see movements like those of plants growing or flowers unfolding in speeded-up or "time-lapse" movies, because the microbes would be growing before our eyes. A bacillus can mature and divide in two every twenty minutes or so. As it did so we would see a constriction at one point as though an invisible string had been drawn tight around it, and the two new bacilli might pull apart a little and whip quickly so as to bring their lengths together rather than their ends. Both from the caviar (cocci) and from the bundles of bacilli, individual beads and rods, and pairs and small clumps, would be continually separating from the main mass, to lie on the stones and dry up and be blown off the planet, or perhaps to be washed away in tiny eddies of sweat and sebum. About as many would be lost this way as were being newly formed, so that the over-all population would stay pretty much the same.

In this region of the scapula on which we landed, we know that some hundreds of thousands (10^5, or multiples of it) of bacteria can be recovered from a square centimeter of clean skin. If we assume that we don't recover more than a tenth or so of what is actually there, we can estimate the resident population as multiples of a million (10^6) per square centimeter. In our new perspective there would be some hundreds per square inch under our feet, but most of these would be under rather than on the surface (as we would see it), and the effect might thus be of a rather thinly seeded lawn, growing, instead of grass, mainly iridescent caviar and crooked sticks, all gently bubbling with the internal movements of growth. Allowing for the microbes detached and drying on the surface scales, and if the proportion of microbes in crypts were—as it may be—more than ten times those on the surface, the impression might be one of a relatively sparse population. It is possible that, seeing it with the composite eyes of Gulliver and Leeuwenhoek, we might compare it with the teeming human population, say, of a city business district at lunch-time or during Christmas shopping, or the beach at Coney Island on a hot Sunday in August—not to speak of the New York subway during rush hour—and decide that the scapula more nearly resembles sub-urbia as seen from the air.

If we could go exploring on man, we would find both more and less densely populated regions. In typically oily skin areas like the wing of the nose, or sweaty ones like the armpit, where detached microbes are not so easily dried and removed, the populations may be ten or more times as high. In the mouth, the throat, and the alimentary tract, plenty of moisture and large stocks of imported food keep the concentrations another tenfold to a hundredfold or even a thousand-fold higher. But within the nose and in the lower reaches of the respiratory tract, on the genitourinary surfaces, and especially in the swift and turbulent tear-stream on the surface of the eye, the popu-lations are much thinner than they are on the shoulder blade, dimin-ishing in some places to almost nothing and, as I have said, becoming nothing at all as we get into areas like the lungs or the bladder.

But in the populated areas where persistent water in microscopic abundance encourages aquatic activity we would find crawlers, swim-mers, darters: animal-like movements added to waving, whipping, and the internal-growth movements we saw on the skin. Nevertheless, the furious, boiling commotion we zoomed down on in Tom Jones's mouth is typical only of the gum crevices and a few other deep cleftlike spaces. Generally speaking, the commonest microbic natives seem con-tent to lie in layers on protected surfaces.

The population densities are highest in the alimentary tract, but even so we have much to learn about them. What we know is based almost entirely on samples we can remove and study out in the labora-tory: of saliva or urine, of fluid sucked up from the stomach or pulled into a syringe from the intestine during surgery, or of feces. The numbers of microbes we recover from these samples must reflect the populations actually inhabiting the surfaces in question, but they may be distorted reflections. Some proliferation of microbes probably takes place only in the food mass, being, that is, purely saprophytic, having no true parasitic connection with our tissues. But not much of it is likely to be so independent of us, or we would not so consistently find the most numerous microbes of feces *only* in feces, never in the food before we eat it. Yet we have no clear idea what relationship may exist between the numbers and proportions of microbes in feces, which we can easily measure, and the numbers and proportions in the crypts and recesses of the gut lining itself. As I mentioned in Chapter 4, studies in animals are beginning to supply this information, but it is still doubtful whether they apply to man. In our stomachs and small intestines, except toward the cecum, the microbic population of the fluid is scanty, but in mice we now know that densities of a hundred million (10^8) or a billion (10^9) bacteria per gram of whole organ are typical. This would represent a population about ten times as dense as that of sweaty or oily human skin areas; these in turn are ten times as thickly settled as dry skin areas.

The numbers of microbes on us are large, but their small individual size makes them aggregate less in bulk than you might imagine. I have already mentioned a high figure for the microbic density of a normal bowel movement: 100 trillion (10^{14}) microbes. A rough calculation based on an alimentary surface area of 1.5 square meters, assuming a true microbic density of 10^9 per square centimeter, would yield a total of 1.5×10^{13}, or 15 trillion microbes. This suggests that we may excrete more than we retain. These numbers are staggering even when compared with the national budget. But estimating that at least 5×10^{11} bacteria, 500 billion, can be packed into 1 cubic centimeter of space like mud, we are dealing with a mere 300 cubic centimeters, or 10 fluid ounces—not much more than a cupful. You may accept this as an estimate of the total microbic population of a man, for consider: His skin area is less than 2 square meters. If we give him an average of 5×10^6 (5 million) bacteria on each square centimeter of his skin, he will have 10^{11} microbes on his outer surface. These would fit into a medium-sized pea and could be slipped into the overflowing cup without being noticed.

My estimates of surface area are conventional and don't take account of the clefts and crypts, the folds and papillae and villi into which skin and alimentary surfaces are thrown, which would make their true surface many times greater. Since the microbes seem to nestle in these very recesses, and evidently tend to get blown or flushed off the free surface toward limbo, the numbers our Gulliver would actually see on the free surface would surely be disproportionately smaller than the true ones. We would seem to be covered not with an unbroken sheet or blanket of microbes but with something like a veil, with much more open space than thread; and everywhere, probably, except for the lowest depths of the alimentary tract, Gulliver would have no trouble seeing our own surfaces through the interstices. When bacteria grow together thickly enough, as they do in cultures, we can see them in aggregate with the unaided eye. But, again excepting the rectum, we cannot see them without help anywhere on skin or mucous membranes.

So I feel impelled to modify a story I told in the first chapter, when we were just starting this journey over populated man. And in doing so I offer apologies to Albert Finney and Joyce Redman, assuring them meanwhile that if I seemed to bring a charge against them it was no very serious one; they are not above mummery to get the effect they want, and I was playing the same kind of game. Nevertheless the charge may not have been true at all, and with all humility I withdraw it herewith.

The point is that we find the picture of seething microbic profusion I described in the gum clefts *only in disease*. The word "disease" covers any recognizable abnormality, and the one in question may be so mild that only a well-trained professional eye would see it, and so common that few people would fail on careful scrutiny to show some trace of it. Found and dredged, it would yield the seething profusion I spoke of, whether it were a localized and utterly trivial spot under a badly made filling or—it isn't always mild!—a frank gingivitis or pyorrhea or acute mouth disease that occasionally becomes major illness. But the healthy young adult with no special dental problems may not show the microbic picture in question at all. One can go scraping around the bottom of the gum clefts of all thirty-two of his teeth and come up with little to show under the microscope: an apparently discouraged spirochete or two will need searching out.

It remains true beyond denial that saliva taken from the same healthy mouth, strained across the teeth from the openings of the salivary glands in the upper cheek and under the tongue, contains some ten million to a billion bacteria in every cubic centimeter. These numbers are made clear by suitable methods of growing the bacteria. Direct examination under the microscope shows disproportionately few, as Leeuwenhoek found. The saliva itself, as it emerges from the glands, contains no bacteria at all. The bacteria are in the mouth. Since most of them are neither in the saliva nor in the free space of the gum clefts, perhaps they are in or on the epithelial scales which saliva contains in plenty, shed in the mouth, like wet dandruff. We know that bacteria can be scraped in great plenty from the corrugated upper surface of the tongue. Maybe most of the microbes in the mouth are attached closely to cells, as they evidently are (in mice, at least) in the stomach and small intestine. Maybe in the mouth as on the skin, they become detached only, as I have said, on their way to limbo. From skin they would be blown or washed off into the great wide world; but from the mouth they usually move toward the gullet, and so perhaps—who knows?—to great adventure.

We need a Gulliver to tell us more, one who could shrink down to half the size of a flea. He would need to survive the terrible earthquakes and tempests of the mouth long enough to see what goes on there and get out again. But his speck, his fourteenth of an inch, would signal his presence, and he would have to find a secure foothold to escape the tides sweeping everything onto the crushing tables of the molars and down into the disintegrating-vats below. With luck he might be fetched by an exploring tongue's tip out to a pair of grasping fingertips, or possibly emerge desperately straddling the point of a toothpick if he escaped being impaled on it. He might have been much too busy warding off his host's violent advances to do much scientific work. With experience and unwavering courage he might learn to enter by stealth while his host slept, maybe by waiting for the portal to open in a sigh or a snore. His report will be eagerly awaited.

Thus are we populated. Microbes cling to us where they can, living in their own way, if they are up to it, off the land and the windfalls. There they grow, usually with a certain moderation, varying with the local climate and especially with the relative abundance of the food supply. While we stay healthy their aggregates resemble the Coney Island crowd or the rush-hour subway jam only at the end of the alimentary trail, where, indeed, they merely accumulate at the gates, awaiting expulsion.

We have every reason to look upon them as part of the environment in which we live. They are the most intimate part of our environment, but they are strictly outside of us; biologists think in these terms of the lining of the gut. Like the alimentary microbes, the food we eat remains outside of us in its course from mouth to anus until our digestive enzymes have broken it into the soluble essences we can absorb. These we take inside, strictly speaking, and use as fuel for our engines and as parts for replacement and new construction. But just as our interaction with food begins outside, so do we interact with our environment in general, and the interaction shapes us and helps to make us what we are. Part of the shaping process is the development of adaptive means to fend off the more unfriendly or dangerous intrusions that make up experience; we become the stronger for learning to handle the hostility of the external world. Our microbes are part of that world.

Even before we knew anything about germ-free animals there was reason to believe we owed something to our normal microbes; but experiments with such animals have removed any grudging doubt we may have felt about the debt, and have made plain some surprising details.

You know that in man as in all mammals the fetus, while its enclosing membranes remain unbroken, is free from microbes as part of its mother's internal environment. Normally it begins to acquire microbes as soon as it makes contact with the outside world. With the aid of exacting aseptic surgery it is possible to remove animals from the uterus just before normal birth, and with a variety of techniques to maintain the animal in a "germ-free" outside environment. In the beginning these techniques were so elaborate and difficult that few laboratories attempted them. But in the last few decades they have been simplified to the degree that many laboratories now work with them, a wide variety of animal species can be obtained and kept germ-free, and germ-free animals can even be purchased commercially for laboratory use and shipped by airplane in germproof packages. The methods, in other words, have become commonplace, and the technical literature has accumulated accordingly.

By seeing how animals get along in the absence of microbes, and then when necessary putting the microbes back to see how they alter things, comparing what we find with otherwise similar "conventional" animals that have had microbes all the time, we learn what the normal microbes do. The most obvious lesson is to abolish at once any notion we might have had that the animal would be generally better off without his germs, springing from the general principle that all germs are our enemies. Far from it. The germ-free animal is, by and large, a miserable creature, seeming at nearly every point to require an artificial substitute for the germs he lacks. He is as a child might be if we could keep him under glass, entirely protected against the buffets of the outside world.

The most striking contribution the normal microbes make to the animal's well-being turns out to be a degree of protection against themselves, somewhat as exposure to wind and cold protects us against both by toughening the skin and insulating us with a layer of fat. But, more generally, the normal microbes are evidently the major stimulus for the whole immunity mechanism, as this is manifested in conventional animals by the development of antibody-forming cells and the appearance of antibodies themselves in the blood. In germ-free animals these producing cells are poorly developed, and antibodies are almost entirely absent. Indeed it seems that the tiny amount of antibody protein (globulin) found in germ-free animals may be a response to the presence in their carefully sterilized diet of dead bacterial cells that are very hard to eliminate. Some of the important antibody-producing substances (antigens) withstand the sterilizing process and elicit the same antibodies as they would have if not so treated.

Antibodies—let me remind you—appear in response to foreign complex molecules like proteins that get inside us, inside in the strict sense of under the skin or the alimentary lining. In the germ-free animal some of these dead bacterial cells, or the intact antigenic molecules in them, must penetrate the gut, evidently in minute amounts, eliciting a rudimentary antibody-forming response. In the conventional microbe-laden animal the same penetration takes place more massively, no doubt starting slowly and building up. Normally there is enough stimulus to get the antibody-forming apparatus into action but not enough to overwhelm the newborn animal: enough to let it develop the cell system in question so that later assaults on a larger scale may be withstood. Presumably some of the microbes that get through in these normal animals are potentially damaging, but the young animal can handle them safely in small doses, and in doing so learns to produce antibody specific for them so that later he can sweep them up much faster and in much bigger dosage. In the presence of microbes, in short, he learns to protect himself against them. In their absence, in the germ-free state the skill lies dormant. It can be awakened later in these animals by injecting antigens or by exposing them judiciously to the normal contaminated environment, sometimes to one bacterial species at a time. The process is not fundamentally different from learning to swim. Something bigger than a bathtub is called for, but the breaking surf is best avoided for a while.

And much as swimming develops muscles that can be used for other activity, the business of learning to keep afloat in a sea of microbes confers other benefits. It is almost as though swimming were our only exercise. Not only does the germ-free animal fail to develop immunity against the normal microbes, but also, since the whole antibody-

forming system fails to develop, the germ-free animal remains susceptible to infection in general, retaining an infantile immaturity toward
the perils of the world.

Germ-free animals of various species have been found to be more
susceptible than conventional animals to experimental infection with
a variety of pathogenic bacteria, at least one protozoan, and several
viruses. Ordinarily we think of such immunity, or the lack of it, as
related specifically to previous experience with the particular microbe or virus; but here the conventional animal is found to have
a degree of protection, in the absence of such specific experience,
which the germ-free animal lacks. In part such protection can be
attributed to relationships between pathogens and normal microbic
species, but some of it is not easily explained this way. For instance,
the colon bacillus can protect an otherwise germ-free guinea pig
against a closely related pathogen, a dysentery bacillus, which produces fatal disease in germ-free animals but has no effect on conventional ones, even though they have had no previous contact with it.
A slightly different, possibly less common, relationship suggests that
environmental microbes we take in during early life—from soil, for
instance—which *do not* survive to become part of our indigenous
population, may nevertheless influence us in this way. Germ-free rats
are much more susceptible to infection with the anthrax bacillus than
conventional ones. This microbe has a host of relatives, none of which
lives with the normal microbes, but they abound in soil and in the
environment generally. The conventional laboratory rat doubtless
comes into contact with these essentially harmless relatives of the
anthrax bacillus and may develop some protection against anthrax as
a result. But it is likely that something more, not yet understood, is
concerned in these evidently non-specific effects. Other instances of
susceptibility of germ-free animals are not explainable at all in such
terms, as with viruses, several of which infect germ-free animals more
easily than they do coventional ones. Absence of the kind of biological
(actually biochemical) relationship that exists between colon and
dysentery bacilli is exemplified in the finding that an accidental contamination of very young germ-free mice with a single bacterial species—a staphylococcus—improved their ability to ward off one of the
large group of viruses, related to polio, named for the town of Coxsackie, New York.

We don't know why this should be so, but it is undeniable that
experience with certain bacteria in early life helps protect animals—

and doubtless man as well—not only against those particular bacteria but against more active pathogens and against certain viruses in addition. There are some exceptions. The rule seems to fail with experimental tuberculosis, where no clear difference between germ-free and conventional animals has been established. And in the instance of amebic dysentery (distinct from the bacillary disease) the presence of intestinal bacteria has been found *necessary* to produce the experimental disease. But these are oddities. On the whole, it is clear that we owe our normal microbes a considerable debt.

Indeed the dimensions of the debt extend well beyond the limits of infection and immunity. In the intestine the presence of a great mass of microbes acts physically, and doubtless chemically as well, to influence the development of the intestinal wall itself; and it is likely that more subtle effects of a similar sort occur wherever the normal microbes are found on our surfaces. The effect of the germ-free state is most obvious in the cecum, at the top of the large intestine. Normally a tough, thick tube, in germ-free animals it is thin and weak and comes to be distended to four to six times its normal size, crowding the abdominal space. The normal contractions of the cecal muscles, which mix the food in it and push it along, become sluggish and feeble. The finer structural details of the intestinal wall also fail to develop, the wall remaining much like that of unborn animals. That it is the absence of normal microbes which induces these deformities is made clear by supplying the animal with microbes. Feeding only colon bacilli to erstwhile germ-free guinea pigs has been found to restore the normal architecture and structural character of the bowel in a few weeks.

There is more to the debt we owe our microbes. In fact there is certainly more to it than has yet been discovered. It has been found, for instance, that the activity of the heart, as measured by the volume of blood pumped by it, is weaker in germ-free than in conventional animals, suggesting that my previous analogy with exercise may be very close. It has been known for a long time that many of the intestinal bacteria can manufacture some of the vitamins we need in our food. Versatile biochemists that they are, they can put together big molecules from little ones sometimes more expertly than we can, just as they can digest and disassemble the big ones in ways that are beyond us. It is the microbes of the gut that help cattle digest the cellulose fiber of grass. Germ-free animals must be fed with very much greater care than conventional animals require, especially with care to supply vitamins in sufficient variety and amount. It looks very much as though the intestinal bacteria can help their host to make up certain deficiencies in his diet, especially of vitamins. Symptoms suggesting certain B-vitamin deficiencies appear in people following treatment for infection with antibiotic drugs which destroy many of the intestinal microbes—an effect attributed in part to loss of such microbic synthesis. Now we have learned that milk cattle, given fodder lacking essential amino acids, have been maintained in good health, with relatively high milk production, the milk having good vitamin content and good flavor—thanks to amino acids synthesized by bacteria in the cow's rumen. Cows, in short—and who is to say only cows?—can actually feed on their intestinal microbes, transforming the microbic substance into their own, some of which becomes their milk. Surely greater love hath no microbe! Recall Hamlet's words to Horatio in the graveyard, leading up to:

> Imperial Caesar, dead and turned to clay,
> Might stop a hole to keep the wind away.
> O, that the earth which kept the world in awe
> Should patch a wall t'expel the winter's flaw!

What magnificent words Shakespeare might have given us if he could have known that animals, man doubtless among them, exchange their substance with such lowly things as microbes! (Shakespeare knew nothing further down the scale than worms.)

We come here to examples of true symbiosis—a mutually beneficial relationship between host (including man) and parasite—which have not yet been demonstrated in such detail as to define a particular microbe among the indigenous population as symbiotic with a particular animal species. But there can be little doubt that the details will be forthcoming. The bifid bacteria in the bowels of nursing infants would qualify if we could be sure the benefits associated with their presence are in fact due to them and not rather to the absence of colon bacilli and other bacteria during the nursing period. The acid-producing bacteria of the young adult vagina would seem to qualify, but we are still not quite sure which of at least two different kinds of bacteria are concerned. The vitamin- and amino-acid-making bacteria would qualify if we could identify them exactly: we know many that *can*, but not the ones that actually *do*; perhaps they work in associations of microbic species—a multiple symbiosis? And here's another example: germ-free piglets cannot digest mother pig's milk; they lack an enzyme that splits the milk sugar (lactose) into glucose, and unless supplied with glucose in their diet (or with friendly bacteria!) they develop severe hypoglycemia and go into general decline.

Knowing things like this, would you willingly separate your infant from his microbes if you could? Or ought you to be glad you can't?

PART II

Taboo, Prejudice, and History

6

Freud's Gold

... civilized men ... are clearly embarrassed by anything that
reminds them too much of their animal origin. They are trying
to emulate the "more perfected angels" in the last scene of
[Goethe's] *Faust*, who complain:

> *Uns bleibt ein Erdenrest*
> *zu tragen peinlich,*
> *und wär' er von Asbest,*
> *er ist nicht reinlich.*

[Literally: "We still have a trace of the Earth, which is distressing
to bear; and although it were of asbestos it is not cleanly."]
  Since, however, they must necessarily remain far removed
from such perfection, men have chosen to evade the predicament
by so far as possible denying the very existence of this inconvenient
"trace of the Earth," by concealing it from one another, and
by withholding from it the attention and care which it might
claim as an integrating component of their essential being. The
wiser course would undoubtedly have been to admit its existence
and to dignify it as much as its nature will allow.

—*Sigmund Freud, Complete Psychological Works*

I do not want to emphasize disease in this book. The core
of our problem is health, a much more difficult subject. The fact is
that we can speak learnedly of disease without knowing where it
begins as a departure from the normal. To define health in negative
terms is to make it an abstraction, of little practical value. Yet disease
in some form is so nearly universal that if we were to find a specimen
of a living thing that seemed entirely free from it, the probing required

57

to establish the fact would inevitably spoil it by inducing disease in it. Physicians are satisfied to define health in a workaday way as a lack of anything in the patient in which they can be interested, whether the basis of their interest is humanistic concern, scientific inquiry, or financial gain. Thus people come to be called healthy whose pains or troubles merely fail to bring out a response in the doctor's eyes or to his fingers or instruments. Or he may pounce on an anomaly that is just what he needs to complete a scientific paper, even though the patient is not bothered by it and is never likely to be. So the patient may object that the doctor's definition is defective in including too much as well as too little. A tendency to define health as a statistical norm, which would give us as a healthy child one with tooth decay, acne, several latent virus diseases, and an incipient taste for murder, is equally unacceptable, at least to me.

Let health be an abstraction, there being little we can do to make it concrete, but let us think of it as an ideal toward which we may hope to aspire with benefit, though we never achieve it. We can then begin to speak of it positively. Let it be a poetic rather than a scientific conception:

> What a piece of work is a man! How noble in reason; how infinite
> in faculty; in form and moving, how express and admirable; in
> action how like an angel; in apprehension how like a god: the
> beauty of the world, the paragon of animals!

This piece of Renaissance affirmation glows like a diamond on black velvet, the more in its context because of the background of suspicion and doubt in which Hamlet embeds it in his speech to his schoolfellows turned spies. The ideal of health has a similar glow as it shines out of the depths of our day-to-day experience with disease. This is the star we follow.

Leeuwenhoek discovered life on man. This was not only his most important discovery: it is one of the most important discoveries ever made by anybody. But ask any schoolboy—ask any American—and you will not escape hearing the opinion that Columbus did something much more impressive. I don't see how anyone can really deny that a thing outside of man, though it be as important to him as his own country or as big as his hemisphere, derives its value from its relationship to *him*; and that he himself, his nature, his composition, his actuality, must in the end be the most important thing he can know. We impart our prejudices with our experience to our children in the process we call education. We teach them about Columbus when they are young and impressionable, and we leave Leeuwenhoek for much later, and only for those who study biology. This is prejudice. The parents of prejudice are ignorance and fear. It is not that Columbus has been soberly adjudged the greater man, but more likely that Leeuwenhoek came at some point to be looked upon as unsuitable for children, and stayed that way. No doubt what happened was that certain adults, like the squeamish Hoole, were repelled by the very qualities in him I think we ought to admire.

Adults are prejudiced; children are less so. Prejudice is learned. Leeuwenhoek's uninhibited pleasure at seeing a new world of microbes in things as forbidding as feces was like the innocent pleasure of an infant discovering the world for himself. Leeuwenhoek was not without prejudice, nor is all prejudice necessarily unhealthy or bad. To Leeuwenhoek "feces" and "dirt" were the same, and a stink was a stink, with no suggestion of enjoyment. But to find a whole world of microbes inside himself and in everyone around him, to discover that man is populated, was cause for great excitement and for wonder and rejoicing. It was a discovery about ourselves, something spectacular; it turned a corner and opened a new vista.

The *significance* of the microbes was irrelevant at the time and has probably always been irrelevant as far as prejudice is concerned. Leeuwenhoek knew nothing of their significance and evidently didn't worry about it. He was excited by the presence of the microbes, their form, their movement, and their numbers, as any explorer would be at finding new and strange forms of life anywhere. They neither attacked him nor rewarded him through his microscopes, and there was no need to fear them or to champion them. In the absence of an overt act, horror, disgust, or fear would arise from ignorance, from some dark and wild supposition. Leeuwenhoek's disgusted gentlewomen, like the cleaning woman who saw her spirochetes through my microscope, needed only an active imagination working in superstitious dread of the dark to make them recoil with repugnance or horror.

Prejudices start being fed to us with mother's milk and with the awareness we learn with our mouths before our eyes can focus sharply. They come out of the wells of love and kindness, and out of the fears and prejudices and the more or less hidden hostility and sadism, of the adults and older children around us. From them we begin to form notions of what is safe and beneficent, warm and pleasant, or of cold, bitter, unpleasant, and dangerous. Mixed with impressions that begin to build a sound foundation on which to walk in the world are others that give us the particular color of our particular culture. The sense of smell, for instance, is conditioned early in man and certainly comes to have much more imitative irrationality in the interpretations it stamps on memory than it does in animals, which do not cover their instincts with the pale cast of thought.

Here we have the problem. Whether or not it would be good to return to infancy and start over without prejudice—and I am not at all sure it would be—we obviously can't do it. We have to live with our prejudices, and all we can hope to do with them is to identify them and try to change the ones we find damaging, or get rid of them. Damage to us is the same as disease; and prejudices might be classified, like microbes, as pathogenic or harmless (or even beneficial). A job of sanitation on the harmful ones would then improve our health, individually and socially. We seem to have accumulated through centuries of ignorance and misdirected learning a freight of garbage which, like the feces and other foul offal with which it deals needs to be faced, however perilous the job may be to our moral eyes and noses, and then disposed of, if possible, rationally.

In this area of human behavior we can turn to Freud for enlightenment. Freud was much less interested in our particular subject than he was in sex, for reasons of unquestioned validity; but he had things to say that seem to me cogent and illuminating in our field as well. What we want from him now is his view of the nature and development of our feelings in relation to the orifices of the body and to their secretions and excretions. It happens that Freud wrote during the early, flourishing era of bacteriology, and I can't help noticing that he seemed virtually untouched by discoveries in the field, which he must have known about as a physician. My earlier suggestion is thereby underscored—that our feelings about feces and saliva have little or nothing to do with the microbes in them and depend on ignorance rather than on knowledge. Freud links all these matters with sex, and whether this preoccupation led him into exaggeration— Freud's principles and practices have not been universally accepted— is unimportant to us. The connection with sex is undeniable, and whether its importance is as determining as Freud thought doesn't matter. It will be clear that Freud's observations in our area are illuminating even if some of his sexual allusions be discounted.

For instance, he says in *A General Introduction to Psychoanalysis*:

. . . infants experience pleasure in the evacuation of urine and the contents of the bowels . . . they soon endeavor to contrive these actions so that the accompanying excitation of the membranes in these erotogenic zones may secure them the maximum possible gratification . . . the outer world first steps in as a hindrance at this point, a hostile force opposed to the child's desire for pleasure—the first hint he receives of external and internal conflicts to be experienced later on. He is not to pass his excretions wherever he likes but at times appointed by other people. To induce him to give up these sources of pleasure he is told that everything connected with these functions is "improper," and must be kept concealed. . . . His own attitude to the excretions is at the outset very different. His own faeces produce no disgust in him; he values them as part of his own body and is unwilling to part with them, he uses them as the first "present" by which he can mark out those people whom he values especially.

I will not attempt to paraphrase Freud. Here are two more quotations, from Volumes IX and XVII respectively, of his *Complete Psychological Works*:

[In early childhood an evacuation] was something which could be talked about in the nursery without shame. The child was still not so distant from his constitutional coprophilic inclinations. There was nothing degraded about coming into the world like a heap of faeces, which had not yet been condemned by feelings of disgust. . . .

. . . a part of the erotism of the pregenital phase . . . becomes available for use in the phase of genital primacy. The baby is regarded as "lumf" [a child's word for feces] . . . as something which becomes detached from the body by passing through the bowel. . . . Linguistic evidence of this identity of baby and faeces is contained in the expression "to *give* some one a baby." For its faeces are the infant's first gift, a part of his body which he will give up only on persuasion by some one he loves, to whom, indeed, he will make a spontaneous gift of it as a token of affection; for, as a rule, infants do not dirty strangers.

Freud notes similar but less intense reactions around urine. And again, from Volume XVII:

Faeces are the child's first *gift*, the first sacrifice on behalf of his affection, a portion of his own body which he is ready to part with, but only for the sake of some one he loves. . . . At a later stage of sexual development faeces take on the meaning of a *baby*. For babies, like faeces, are born [in the child's view] through the anus. The "gift" meaning of faeces readily admits of this transformation. It is a common usage to speak of a baby as a "gift." The more frequent expression is that the woman has "given" the man a baby; but in the usage of the unconscious equal attention is justly paid to the other aspect of the relation, namely, to the woman having "received" the baby as a gift from the man.

(In the opening pages of *Tristram Shandy*, Sterne mentions the quaint notion of a "homunculus"—a tiny human being completely formed, deposited by a man in the womb of his mate, leaving to the mother only the role of nurturing the little fellow until he is big enough to come out into the world.)

Freud returns to the infant's attitude toward feces in his Introduction to the German translation of Bourke's *Scatologic Rites of All Nations*, a book we shall come back to. This is the source of the quotation at the head of this chapter; and here we also read:

> . . . the chief finding from psychoanalytic research [on excretory functions] has been the fact that the human infant is obliged to recapitulate during the early part of his development the changes in the attitude of the human race towards excremental matters which probably had their start when *Homo sapiens* first raised himself off Mother Earth. In the earliest years of infancy there is as yet no trace of shame about the excretory functions or of disgust at excreta. Small children show great interest in these, just as they do in others of their bodily secretions; they like occupying themselves with them, and can derive many kinds of pleasure from doing so. Excreta, regarded as parts of a child's own body and as products of his own organism, have a share in the esteem—the narcissistic esteem, as we should call it—with which he regards everything relating to his self. Children are, indeed, proud of their own excretions and make use of them to help in asserting themselves against adults. Under the influence of its upbringing, the child's coprophilic instincts and inclinations gradually succumb to repression; it learns to keep them secret, to be ashamed of them, and to feel disgust at their objects. Strictly speaking, however, the disgust never goes so far as to apply to a child's own excretions, but is content with repudiating them when they are the products of other people. The interest which has hitherto been attached to excrement is carried over on to other ob-

jects—for instance, from faeces on to money, which is, of course, late in acquiring significance for children.

Freud speaks of the relationship of feces to money in several places. For instance, in Volume IX, in a discussion of a neurosis in which the patient is "exceptionally *orderly, parsimonious,* and *obstinate,*" which Freud calls "anal erotism," he goes on to say:

> Cleanliness, orderliness and trustworthiness give exactly the impression of a reaction-formation against an interest in what is unclean and disturbing and should not be part of the body . . . It might be supposed that the neurosis is here only following an indication of common usage in speech, which calls a person who keeps too careful a hold on his money "dirty" or "filthy." But this explanation would be far too superficial. In reality, wherever archaic modes of thought have predominated or persist—in the ancient civilizations, in myths, fairy tales and superstitions, in unconscious thinking, in dreams and in neuroses—money is brought into the most intimate relationship with dirt. . . . It is possible that the contrast between the most precious substance known to men and the most worthless, which they reject as waste matter . . . , has led to this specific identification of gold with faeces.

Freud's observations were part of his lifelong study of mental or emotional illness, which he traces to repression or distortion of feelings concerned with natural functions, in particular the sexual functions. Aside from his lack of interest in microbes, he was only incidentally concerned with what interests me in this context, the source and origin of prejudices that relate to these functions. What appears from Freud's work is that two separate roots of these prejudices can be discerned. These are, first, the sexual-excretory connection, which Freud appropriately calls "cloacal"; and, second, the much less clearly defined idea of defilement. The second root grows into relation with both sexual and excretory functions, but seems more closely bound to the latter; yet the question hovers over it—and Freud is not concerned with this at all—does it have any rational basis? We must put this question off to the next chapter and look in other sources for an answer.

Freud speaks of matters that concern us other than feces, and as our own inescapable revulsion diminishes in moving to them, it may be easier to accept the sexual connection. In Volume X of his *Complete Psychological Works,* Freud says in passing:

... I should like to raise the general question whether the atrophy of the sense of smell (which was an inevitable result of man's assumption of an erect posture) and the consequent organic repression of his pleasure in smell may not have had a considerable share in the origin of his susceptibility to nervous disease. . . . For we have long known the intimate connection . . . between the sexual instinct and the function of the olfactory organ.

The following passage in *The Basic Writings of Sigmund Freud* (Brill) feeds our doubt that there is anything rational about our revulsion:

The employment of the mouth as a sexual organ is considered as a perversion if the lips (tongue) of the one are brought into contact with the genitals of the other, but not when the mucous membrane of the lips of both touch each other. In the latter exception we find the connection with the normal. He who abhors the former as perversions, though since antiquity these have been common practices among mankind, yields to a distinct *feeling of loathing* which restrains him from adopting such sexual aims. The limit of such loathing is frequently purely conventional: he who kisses fervently the lips of a pretty girl will perhaps be able to use her tooth-brush only with a sense of loathing, although there is no reason to assume that his own oral cavity for which he entertains no loathing is cleaner than that of the girl . . .

Speaking as a bacteriologist with some special competence in the areas mentioned, I am quite unable to object to this. A little farther down on the same page we find this:

It is even more obvious than in the former case, that it is loathing which stamps as a perversion the use of the anus as a sexual aim. But it should not be interpreted as espousing a cause when I observe that the basis of this loathing—namely, that this part of the body serves for the excretion and comes into contact with the loathsome excrement—is not more plausible than the basis which hysterical girls have for the disgust which they entertain for the male genital because it serves for urination . . .

A similar comment is made by Freud in Volume VII of his *Complete Psychological Works*. Speaking of "displacement" in a case of hysteria, manifested as disgust in a girl following a kiss and embrace, he says:

. . . the feelings of disgust . . . seem originally to be a reaction to the smell (and afterwards also to the sight) of excrement. But the genitals can act as a reminder of the excretory functions; and this applies especially to the male member, for that organ performs the function of micturition as well as the sexual function. Indeed, the function of micturition is the earlier known of the two, and the *only* one known during the pre-sexual period. Thus it happens that disgust becomes one of the means of affective expression in the sphere of sexual life. The Early Christian Father's *"inter urinas et faeces nascimur"* [we are born between urine and feces] clings to sexual life and cannot be detached from it . . .

This is as close as Freud gets to the notion of defilement, the origins of which are the concern of the next chapter. Be it noted that the Early Christian Father had, it seems, more than a casual or incidental part in the story.

7

How Did We Go Wrong?

The Hebrews [were] the first Puritans in their nice observance
of law, particularly of all laws relating to exposure of the body,
the sexual and other natural functions.

—Reginald Reynolds, *Cleanliness and Godliness*

Biologists have shown us that babies start life in the womb
as a single cell and grow through a sequence of stages that roughly
recapitulate the long evolution of the human species. It may be
assumed that they are born without any prejudices; and, while there
is no necessary parallel between biological and cultural development,
it need not come as a surprise that early man lacked the prejudices he later turned against himself. These assumptions are more or
less self-evident, and the facts tend to support them. But we find
prejudices developing very early, long before Leeuwenhoek, long before microbes could even have been dreamed of. They must have been
based on something else.

A search for the origins of man's prejudices about himself shows,
as I need hardly tell you, that mystical ideas, myths, superstitions,
taboos, and the other furniture of the pre-scientific mind must have
begun to appear soon after the animal we call *Homo sapiens* first
emerged on the earth. Yet the ideas we are looking for did not show
themselves in recognizable form until much later; and while we may
assume again a long, slow evolution from early roots, the plant itself
and its typical bitter fruit are clearly phenomena of the Christian era.
The record is not easy to disentangle, partly because its strands tend
to be strongly colored with the very prejudices we are trying to separate from the substance under them. If the material we are looking

66

for was not simply omitted in the first place, it is likely to have been expurgated by a later observer, or at least distorted or covered over with euphemisms.

The ideas we seek are those man entertained about his body, its parts, functions, and products; let us lump them together for simplicity's sake as ideas about excretion. Such ideas, and ideas of sex, with which they were presumably associated from the start, quickly became subjects of magical interpretations and practices, that is, speaking broadly, of religion. But whereas sex and religion are matters that quicken man's thoughts today and must always have done so, ideas about excretion, whatever men may once have thought of it—and it may well have been less interesting from the start than the other two—have tended to be suppressed by the very prejudices whose origins are our immediate objective. Yet, with due allowance for these difficulties, it is possible to sketch the broad outlines of the evolution of excretory prejudice.

The earliest records bearing on the subject are the carvings, sculptures, and cave paintings of southern France and northern Spain, dating from the early Upper Paleolithic period, some 30,000 years ago. In the art itself, which is typified by the cave paintings at Lascaux in France, is clear evidence that man had reached by then a level of intellectual development quite equal to our own. This circumstance is hardly surprising, since 30,000 years is a brief interval of geologic time. As far as I know, this earliest record contains nothing representative or suggestive of excretion, but sex is abundant and striking. Religion is implicit in the very existence of this art, as well as in its consistent content and structure. The repeating patterns in which brilliantly naturalistic representations, especially of animals, occur together with highly stylized abstractions and geometrical figures that must be symbolic, testify to widespread religious practice. The record as it has been exposed by modern observers suggests a consistent development of a pattern of ideas over some twenty millennia, stopping short of the beginnings of what we call history.

Sex is unmistakable in the naturalistic art, and recent studies have identified the abstractions and geometric figures as male and female symbols. In the former the genitals of both animals and human beings tend to be emphasized. The vulva is often exaggerated in statuettes or figurines ("Venuses") representing pregnant women with huge breasts but showing little or no detail of the face. In the male the penis is often shown erect. What all this means is not entirely clear; but there is nothing like a fig leaf, nothing that might suggest the root of the modern notion of shame.

Twenty-five thousand years or so after Lascaux, when man's cultural development had progressed apace and the accumulating records had become literary as well as graphic, there is still little evidence of restraint on expression. But now we find the path strewn with the distortions imposed by later scholars—of which the fig leaf itself is a sort of symbol—and it becomes necessary to pick our way among them with caution.

It is well known that the Bible is much franker in speaking of both excretion and sex than we tend to be; and scholarly efforts to disguise this fact were already being revealed (and laughed at) in Shakespeare's time. Sir John Harington, the Elizabethan poet and inventor (whom we shall meet again), poring over one of the English Bibles that preceded the Authorized Version of King James, noticed that a phrase that appears twice in the Old Testament (I Samuel 24:3, Judges 3:24), "covering the feet," refers in fact to evacuating the bowels. The corresponding phrase in the Latin Vulgate edition is *ut purgaret ventrem.* Where the King James Version in Judges says, "Ehud . . . covereth his feet in his summer chamber," the Vulgate says, *"Forsitan purgat alvum in aestivo cubiculo."* The sense is made clear in later Bibles by marginal glosses, in the phrases "it came out at the fundament"; and, specifically directed to the words "covereth his feet," the gloss reads "doth his easement."

Harington—let me note in passing—also mentions the word "draught" in the New Testament (Matthew 15:17, Mark 7:19), where it clearly means "privy." This is one of many excremental terms that are strange to our view but were evidently in common use by Elizabethans, including Shakespeare. It appears with a suggestion of modern gutter speech in the words of the scurrilous Thersites toward the end of *Troilus and Cressida.* Hector is in the Greek camp preparing for the battle with Achilles, and exchanges a courtesy with the enemy:

HECTOR: Good night, sweet Lord Menelaus.
THERSITES: Sweet draught! "Sweet," quoth 'a! Sweet sink, sweet sewer.

Some examples of the sort of distortion I am referring to are given in the following quotation from Hyde's *History of Pornography:*

. . . the great English Greek scholar Dr. Gilbert Murray always insisted upon translating the verb "to break wind" as "to blow one's nose." We first hear mention of a suggested literary censorship with a view to suppressing possible pornography toward the end of the fourth century b.c., when Plato proposed to do for Homer what Bowdler succeeded in doing for Shakespeare, namely to expurgate the author's words for the use of the young.

One wonders, indeed, how far back such practices go. Notions of magic probably go back to the very beginning. Words have always tended to have magical overtones, and older people must always have tried to protect the young against their imagined evils. But we know that in Greek and Roman times there was a freedom of literary and artistic expression that was later curtailed. We can still use the words derived from their languages, and the genitals that were not chopped off by later apostles of purity are still intact under the censor's fig leaves. Add this to the evidence of the Old Testament, and supplement it with what we know of ancient civilizations in India and in both Near and Far East; put in with this what we are told of pagan practices generally and what has always been in plain sight in the barnyard—and we can begin to get the field of view in focus. Herodotus in his *History* suggests the early roots of the problem by telling us that the Egyptians

> eat their food out of doors, but retire for private purposes to their houses, giving as a reason that what is unseemly, but necessary, ought to be done in secret, but what has nothing unseemly about it, should be done openly.

But, recognizing these antecedents, we may nevertheless date the origin of the problem in its modern form in Rome at a time no earlier than the second century a.d., when Juvenal was setting down word images that later came to be called indecent.

The particular prejudices we are searching out began around that time, and their immediate roots go back through our Jewish-Christian forebears. Still earlier sources can be seen in the complex of fertility-sex worship and the rites of sacrifice and ablution of pre-Israelite and early Greek polytheism, out of which sprang the notions of hygiene characteristic of our particular culture. Somewhere along the way a connection appeared between a feeling for cleanliness among these early peoples and a sense, perhaps passing through modesty into shame pertaining to excretion, and extending into feelings about sex, and so growing into the germinal root of this whole set of prejudices. But the process is one of evolution, and no sharp point of origin need be looked for.

The full development of the sense of shame, however, in the sense of guilt feelings pertaining to functions and emotions shared by everybody, is evidently a heritage of the Christian Church. It blossomed in the teachings of the early Christian Fathers, especially Saint Paul and his disciples, was promulgated by Augustine in the fourth century A.D., and reached its height during the Reformation. Abnegation was its essence, the wickedness of carnal pleasure its basis. Salvation in the life to come depended on denial of the body on earth.

Harry Elmer Barnes tells us that Saint Augustine

> left a potent heritage to European thinkers: his violent reaction against sexual indulgence. He had . . . in his youth led a wild and dissolute life, judged by Christian standards, and he had been loath to abandon this life. His prayer on this point—"O Lord, make me chaste, but not quite yet!" is a classic. But when he was finally converted his conduct and attitude in such matters changed violently. He became very bitter on the matter of sex and denounced it. He suffered acutely from what modern psychologists call "overcompensation." He traced all human ills to sexual indulgence and redefined original sin in such terms. Original sin was thus portrayed as the origin of sexual intercourse, the blame for which was thrown upon Eve. These morbid eccentricities of Augustinian thought, growing out of his own erratic personal experience, were able to pervert human thinking on sexual matters for a thousand years, and their influence is still strong with millions. Augustine's personal sex neurosis was elevated to a dominant position in western European ethical theory.

The specifically excremental element in Christian abnegation was evidently contributed by Martin Luther, who, as Norman O. Brown tells us in his *Life against Death*, quoting Luther's own words, not only received his knowledge of the "justice of God" while seated "on the Privy in the tower" of the Wittenberg monastery, but went on to combine current folklore and theological speculation with his own experience to relate the image of the Devil to feces and associated vividly scatological images.

During the Renaissance open conflict broke out between liberating forces seeking affirmation and complete freedom of expression, and repression from both the Roman and the rising Protestant churches. The repressive forces eventually prevailed, especially—or from my viewpoint, at all events, distinctively—in the English-speaking world, under the guise of Puritanism. Before this happened there had been, of course, the greatest efflorescence of art since the Greeks. Science as we know it today was also born during this period. In part the efforts of the Church were directed squarely against these major Renaissance forces, notably those of literature and the theater. But the intellectual liberating movement must have suffered equally, if not more, by reflection from the violence of the Church in its war against the remnants of pre-Christian practices, a war symbolized in the great witch-burnings of the period, in which the work of the Catholic Inquisition, both papal and Spanish, is hardly to be distinguished from that of the Puritans in both old and New England. With the less bloody but hardly less powerful later influence of Victorianism, we can trace an unbroken line of bigoted suppression of man's natural impulses down to the present day. Its central theme has been a neurotic denial of the pleasure of sex, but associated with this has been an equally perverse effort to suppress the normal human feelings associated with excretion and other functions proper to my theme. Counterforces have appeared in scientific rationality, notably in Freud, and, as before, in a groundswell of expression striving to be free, both artistic and heretical. It looks as though the forces of liberation may prevail this time. But as the Church has lost much of its early perverse potency, the state has taken over part of its function, and we have a new perversity to contend with in the marketplace.

The historical record of the development of these prejudices is supplemented by an indirect record based on the study of contemporary peoples with cultures different from our own, especially those that seem by our standards to be primitive and thus to be counterparts of man as he existed in past ages. But the observer, like the rest of us, is unavoidably prejudiced in some degree; and the record as he gives it to us is accordingly filtered through a mind that selects and rejects in accordance with its background. Anthropologists can hardly be expected to study aborigines without a trace of invidious comparison with themselves. This bias is startlingly evident in the special area discussed in this book, and is further exaggerated here by the coincidence that my principal sources come from the Victorian period. The result is at least twofold. Either the specific subject-matter I am looking for is glossed over or omitted, or else it is given with gestures of apology that, to me at least, reflect more on their source than on their subject. Captain Bourke, for instance, whose treatise on scatology (we have looked into Freud's introduction to it) is invaluable for my purposes, gives the impression in his opening pages—perhaps intentionally, with tongue in cheek—of having written the book with one hand while holding his nose with the other. Many of his sources show the same comical posture without subterfuge.

The great anthropologist Sir James George Frazer is himself not above this criticism, and certainly omitted from *The Golden Bough*, out of Victorian delicacy, much that he must have found and that would have been useful to us. Later anthropologists have tended to follow him in this practice. There is nevertheless much of immediate value to my theme in Frazer's work. He offers a forthright definition of magic as "a spurious system of natural law as well as a fallacious guide to conduct; it is false science as well as an abortive art." Upon this idea is developed the fruitful concept of *sympathetic* magic, in which parts or representations of a person, including drawings, images, and even names, could be substituted for the person himself in order to cure him or to do him injury, intentionally or not. Among such parts of persons, Frazer speaks at different points of "nails, eyebrows, spittle, and so forth," and of hair, sweat, and even breath; but he seldom refers to grosser things except indirectly. This section is followed by one on tabooed foods, in which the author's avoidance of any mention of excrement could hardly be plainer. A chapter on "Tabooed Words" notes the savage's inability "to discriminate clearly between words and things" and is then concerned entirely with taboos on names of persons.

Frazer speaks in several places of practices bearing on sanitation, and again the text is useful both for what it says and for what it leaves out. The following suggests how a useful practice may have had an irrational origin:

> The superstitious fear of the magic that may be wrought on a man through the leavings of his food has had the beneficial effect of inducing many savages to destroy refuse which, if left to rot, might through its corruption have proved a real, not merely imaginary, source of disease and death. Nor is it only the sanitary condition of a tribe which has benefited by this superstition; curiously enough the same baseless dread, the same false notion of causation, has indirectly strengthened the moral bonds of hospitality, honour, and good faith among men who entertain it. For

it is obvious that no one who intends to harm a man by working magic on the refuse of his food will himself partake of that food, because if he did so he would, on the principle of sympathetic magic, suffer equally with his enemy from any injury done to the refuse. This is the idea which in primitive society lends sanctity to the bond produced by eating together; by participation in the same food two men give, as it were, hostages for their good behavior . . . The covenant formed by eating together is less solemn and durable than the covenant formed by transfusing the blood of the covenanting parties into each other's veins, for this transfusion seems to knit them together for life.

Another of Frazer's concepts that bears on sanitary practices is the failure of primitive peoples, and indeed of pre-Christian peoples including the Jews and the Greeks, to distinguish clearly between holiness and pollution. To the savage, one who was pronounced unclean or polluted shared with one conceived of as holy the common features of being dangerous and in danger. Since the Jews could neither eat swine nor kill them, it seemed uncertain whether the animals were objects of worship or abomination. Jews washed their hands after touching a pig, and also after contact with a holy object or after reading the scriptures. In the Greek ritual of sacrifice, the sacrificer was not to touch the offering, and after it was made was required to wash his body and his clothes before he could enter a city or his own house. Frazer describes analogous washing and cleansing rites among the Creek Indians of North America during the midsummer festival of the first fruits, the chief ceremony of the year, associated with a fast followed by "the devotee drinking a bitter decoction of button-snake root 'in order to vomit and purge their sinful bodies.'"

It was hardly more than by accident, then—by a lucky shot in nearly complete darkness—that ancient practices we think of as hygienic, or others that suggest a certain prescience, like the dietary laws of the Jews, managed to yield beneficial results. Some gain accrued to experience, but generally the method was as likely to kill as to cure. Control and prevention of disease, the goal of hygiene today, were almost entirely hit or miss before Pasteur. When he and Robert Koch and the other early bacteriologists showed that disease may have specific causes in the form of microbes, and that finding and destroying these microbes, or immunizing specifically against them, stopped the disease in question, they introduced as great a revolution in man's conception of himself as Darwin did when he pointed to our descent, as he called it, from other species.

Darwinism came a little earlier, and perhaps for that reason, and because it seemed to undermine the Adam-and-Eve myth directly, and maybe even because it was theoretical and therefore easily argued with, it blew up more of a storm among the bigots than the germ theory did. Thomas Henry Huxley's carefully marshaled facts and brilliant words explaining and defending Darwin could not match the clinching public demonstration by Pasteur that his vaccine could save the lives of animals inoculated with anthrax, the unvaccinated dying before the observers' eyes.

Together these two events were as climactic a show of the constructive power of science as the atomic bombs of August 1945 were a proof of its capacity for devastation. What Darwin and Pasteur taught us—that man is a member of the living community, no more and no less, and that there is reality in cause-and-effect, with knowledge leading to control—ought to have done more for us than in fact it did. From them we ought to have learned to throw away the "spurious system of natural law" and the "fallacious guide to conduct" of the old magic. We did, of course, begin to make headway against the destructive forces of the environment. We ought to have made much more.

If children had been taught in Pasteur's day to understand what had happened in the world—and obviously they were not—magic would have evaporated in a generation, and the prejudice that saturated it would have crystallized in plain view and could have been dissolved in rationality. We are trying to do just this today, but it might have been easier then, before science had moved so far beyond the common understanding as to rekindle a belief in magic —before its forces spread themselves so widely, with reality and fantasy almost hopelessly intermingled, in the exploration of the infinitely distant and the infinitesimally close.

The ever-accelerating pace of science, piling wonder upon wonder, goes on before the eyes of all who watch television or read the popular magazines. Yesterday's impossible is today's commonplace, and it seems hard to go wrong in predicting tomorrow. No homely thing like microbes delving into our healthy skin and guts is necessarily beyond change; yet this particular homely thing, unless I am vastly mistaken, shows no sign of bending under the force of modern technology. I think we will continue to live with our microbes, and, if so, it is going to be necessary to deal with the prejudices that surround them, or we will go on suffering their consequences.

Dealing with them is the job in hand. Let us assume that it is not too late to evaporate off the magic and get a good look at what is underneath, at the stuff the old prejudices have buried, pushed aside, distorted. The stuff is what I have lumped together as ideas of excretion, often called scatology. Part of it is spoken of as obscenity. The first point of attack on the problem may be verbal, since obscenity is largely a matter of words. Can it be a question of words having power to damage you, or to stunt the growth of children? If so, would this not be sympathetic magic?

At this point we need to take a good hard look at some ordinary little words, and at what it is that makes us prefer some words and avoid others.

8

Don't Mention It

We have taught Ladies to blush only by hearing that named
which they nothing fear to do. We dare not call our members by
their proper names, and fear not to employ them in all kinds
of dissoluteness. Ceremony forbids us by words to express lawful
and natural things; and we believe it. Reason willeth us to do
no bad or unlawful things, and no man giveth credit unto it.
Here I find myself entangled in the laws of Ceremony, for it
neither allows a man to speak ill or good of himself.

—Montaigne, *Of Presumption*

Wₑ grow into a world of custom and become inured to
it so gradually that although we are aware of its irrationality whenever we happen to think of it, we seldom bother. The logic we see in
children is precisely a freedom not yet shackled by custom. It is part
of childish charm; we speak of it as innocence, and the very custom
that binds us leads us to apply the bonds to children as an act of
love. Love is itself irrational, of course, and that seems to make
everything right.

As we grow toward maturity we learn that certain natural actions
are incorrect in polite society and that certain words must not be
spoken even though both actions and words are known to everybody.
Some of these customs have been hardening even during my memory
here in the United States, as ordinary people take on the ways
formerly reserved to the upper classes. Even children may no longer
urinate outdoors in the presence of strangers, although it is still
correct for a stylish lady to oversee the same behavior in her dog with
the tender affection otherwise reserved for the nursery. By a training
process that encourages evasion and euphemism we learn not to speak

76

directly of such things, with socially acceptable variations depending on class and sex and situation.

Both the actions and the words have come to be regulated not only by custom but by law, fortified with condign punishments. The actions are natural functions, and when the force that restrains them in one place lets them out at another we speak of perversions, and, confusing effect with cause, we are all the more convinced of their impropriety. Politeness allows us to keep urethral and anal sphincters tightly closed, but only so long before the impulse to release the pent-up liquid, solid, or gas becomes uncontrollable. For the itch, the sneeze, the cough, necessity permits propriety to be tempered, and indulgence is mannerly. We appeal to hygiene without logic. The audience's coughing may drown out the music or the action, and if the affront to the ears is politely more acceptable than that to the nose from the effusions of other body orifices, the later consequences in terms of the common cold and influenza more than make up the difference.

The most powerful of these natural impulses being sexual, it is they rather than the excretory ones that set the pattern of custom and law. Yet it is hard to know why this should be so. It seems irrelevant that the sexual drives develop later than the others and, unlike those others, lose some of their force with advancing age. Nor that social pressure is able with one set but not with the other, in extreme instances, to bottle them up entirely, although with inevitable consequences now accepted as unnatural or pathological. Does the reproductive function entail higher stakes for society than the excretory ones? In the beginning fertility was concerned in different ways with both, and any such distinction ultimately looks forced. The reason for the focus of repression on sex takes shape in a superlative irrationality: the most intense of our pleasures comes to be forbidden.

In the long run repression only postpones or perverts natural actions, and the same is true for words: in the end the result is not the one intended. The law finds it difficult to deal with the actions except at levels relatively mild or petty. Shakespeare wove a whole play, *Measure for Measure*, around the futility and hypocrisy of sexual repression. And characteristically he put the wisest words into the mouth of a lowly character:

POMPEY: Does your worship mean to geld and splay all the youth of the city?
ESCALUS: No, Pompey.
POMPEY: Truly, sir, in my poor opinion, they will to't then.

It is not much easier to enforce laws against words, especially written words. In Jewish and Christian history, laws of church and state, combined or separately, and customs associated with them, have forbidden the utterance of words in two general categories, *blasphemy* and *obscenity*. The first begins with the Mosaic injunction against taking the name of the Lord in vain, but includes the king as God's representative. Hence *cursing* and *swearing* are directly forbidden. More generally banned is *profanity*, which, originally opposed to *holiness*, has come to cover a range of meanings from irreverence to any contemptuous or "vulgar" utterance, with overtones of pollution or defilement. These meanings merge with the ones included under *obscenity*, but this is a profane rather than a sacred notion, having been enforced by the state after temporal power was taken over from the Church. In its narrower meaning, as a legal concept, "obscene" is defined as "offensive to chastity or to modesty." That both defining words have local, relative, or flexible meanings is suggested by the relatively free love, as anthropologist Margaret Mead described it, among adolescents in Samoa (whom we disparage accordingly as "savages"), and by such Japanese practices as public bathing in the nude. Obscenity includes *pornography* and *scatology*, which deal respectively with sex and with excrement.

Searching further for the sense of these English words, we find a group of others that are defined in terms of one another, with varying shades of meaning. If there is a fixed point in any of their definitions it is the words *obscene* or *obscenity*, for which each of them can be a variant. The words are *dirty, smutty, filthy,* and their corresponding nouns, as well as *foul* and *nasty*; and—a branch into another notion— *rude, coarse, gross, indecent,* and *vulgar*. The first group covers the general idea of pollution or defilement, while the second, in the root meanings of the words or in usage, speaks to distinguishing marks of the lower classes as compared with the wellborn or "gentle." Both groups of meanings can be identified as mythical by a simple chemical test: exposed to strong light, the meaning evaporates, leaving no residue.

The idea of pollution or defilement is clearly traceable back to sympathetic magic, even to the degree of its entanglement with holiness. It preceded precise knowledge of hygiene, having no more connection with private or public health practice than it has with the science of microbes. Pollution is very much a fact of life today, but its source is not man's biological products or functions, still less his language, but rather his more sophisticated practices and creations, including cigarettes, insecticides, chemical wastes, and the excretions of the automobile, not to speak of materials and devices fashioned to destroy man and his works in war.

Obscenity defined as *vulgarity* (Latin *vulgus*, the multitude, the common people) is in practice clearly a matter of "lower-class" speech. *Obscenity* and *pornography*, as the two words are interchanged in legal practice, are both defined in terms of the use of little words of common speech of Middle or Old English derivation, hence going back to ancient Britain before the Norman invasion. Not any early English words, of course, but specifically those relating to matters sexual or excremental, or more generally to body parts, functions, and products. In fact, the rest of this vocabulary—words of this vintage not under the ban—far from being frowned on, is actually favored in modern speech and writing by all the better guides to English style. Clear and direct expression is based on the use of short and therefore emphatic old English words in preference to long ones of Greek or Latin origin, which tend to be pretentious if not merely ornamental and meaningless.

Such direct English speech would be a reversion to the vulgar if that word retained its original meaning, but as a result of an about-face flouting etymology, frills and foolish or self-conscious decoration of speech, once the hallmark of the aristocracy, are now themselves called vulgar. But the simple, direct Anglo-Saxon monosyllables for sex and excrement are still forbidden—not all, or always, and the rule may perhaps be bent in the proper mouths. Winston Churchill, having need of dramatic clarity, spoke of "blood, sweat, and tears." The word *sweat*, both as noun and as verb, is preferred in the language of science; but in the commercials on television it is still *perspiration*. Frazer preferred *spittle* to *spit* as a noun; the second is hardly replaceable as a verb. But to dentists the fluid itself is never anything but saliva. (For that matter few dentists seem to be able to use the Anglo-Saxon word *mouth* in formal speech. Even to the patient they are likely to say "rinse out" instead of "rinse your mouth"; and to their peers, in proper speech or writing, it becomes "oral cavity" or even "buccal cavity," and somehow a purification is implied. I remember a rather eminent woman pediatrician who seemed to block on the word "children" and invariably went round about with something like "the younger age group." Such efforts to make words pretty are of course not confined to forbidden ones.)

Such behavior, in any case, is called snobbery. The snob chooses his words to convey the idea that he belongs—or would like to belong —to a class above his audience. He is in fact telling them where to get off, or literally, putting them down. And when it comes to words for body parts, functions, and products he enjoys a double advantage. The lowly folk are forbidden to use the simple words they know, while he is free to use the Greek- and Latin-root words they don't know. Robert Graves tells the old story:

> The soldier, shot through the buttocks at Loos, who was asked by a visitor where he had been wounded, could only reply, "I'm sorry, ma'am, I don't know: I never learned Latin."

Since class distinctions and associated snobbery are universal among peoples everywhere, one would expect word distinctions to be universal too. But there seem to be wide variations among different peoples. As the heritage of Puritanism came to dominance under the aegis of the Anglican Church, so it also flowered into speech forms that seem peculiar to English-speaking peoples, or at least developed among them in its most extreme form. An example that rings true to me, even though its author's tongue never left his cheek as he wrote it, is given by Norman Douglas, best known for *South Wind*, in the introduction to his little book of limericks. Speaking of the reaction to what he calls the Puritanism of Savonarola, he says:

> A self-respecting Florentine would consider his life ill-spent had he not tried to add at least one blasphemy of his own composition to the city stock . . .

And he goes on to suggest that the Englishman's parallel form of artistic creation is the ("obscene") limerick, to which the Frenchman or the Spaniard reacts either with dazed bewilderment or with disapproval: "They regard these things as dirty." In contrast it appears that to a Frenchman the word *merde*, whose English equivalent is a red flag to the Comstock bull, is almost as freely available as are its euphemisms in our language, such as *pshaw*, *shucks*, and the now forgotten Elizabethan *skite*. That such freedom is not peculiar to Latins is suggested by an apparently similar free use of the German word *dreck*, which showed up not long ago in a book review in that great arbiter of English usage in this country, *The New York Times*.

A scholarly approach to this question appears in the little book from which I have already quoted, *Lars Porsena, or the Future of Swearing and Improper Language*, by the English poet, novelist, and classical scholar Robert Graves. The title comes from the first verse of one of Macaulay's *Lays of Ancient Rome* ("Horatio"):

> Lars Porsena of Clusium
>   By the Nine Gods he swore
> That the great house of Tarquin
>   Should suffer wrong no more.
> By the Nine Gods he swore it,
>   And named a trysting day,
> And bade his messengers ride forth,
> East and west and south and north,
>   To summon his array.

Graves's main thesis is that as the number and potency of our gods have declined, so the force and ingenuity of swearing have dwindled. He focuses on swearing, or blasphemy, rather than on the broader area he calls "improper language," and on his selected topic his treatment seems to me definitive. He traces a general deterioration in the force of the oath, originally a rebellion against the temporal power of the Church, to the loss of that power and further:

> The triumph of Protestantism is, perhaps, best shown by the decline into vapidity of "By George!", the proudest oath an Englishman could once swear; for the fact is we have lost all interest in our Patron Saint . . .

He says in passing:

> Public reference to a man's navel, thighs, or armpits, even, is a serious affront; from which the size of the "breeches of fig leaves" tailored in Eden may be deduced. It is difficult to determine how far this taboo is governed by the sense of reverence, and how far the feeling is one of disgust and Puritanic self-hate. But in any case, the double function of the taboo'd organs, the progenitive and excretory principles, has confused the grammatic mind of civilization . . .

The "serious affront" is mitigated by poetic license. "Ogres and Pygmies," a poem by Robert Graves, begins:

> Those famous men of old, the Ogres—
> They had long beards and stinking armpits.

Or perhaps, if the poem came after *Lars Porsena* (which appeared in 1927, when Graves was thirty-two), it represents a change of mind. Even so the book is a courageous one, not fearing to challenge convention; and such courage seems to have characterized Graves's work generally, down to his new translation of the "Rubáiyát of Omar Khayyám" in 1968. He has in fact been attacked by scholars as an iconoclast; he suggested that Jesus survived the Crucifixion; that the Roman Emperor Claudius, far from being a despot, was a mild scholarly fellow; and that it was one of Homer's children, not Homer himself, who wrote the *Odyssey*. But I am impelled to criticize *Lars Porsena* for the opposite reason: it is not iconoclastic enough for me. For all its challenge to convention, it stays too close to the bounds of British gentility. Graves can laugh that:

> The lavatory-taboo still survives with us at meal-times, but we find it difficult to understand the extraordinary customs to which the morbid enlargement of this natural reserve led. For instance the playwright Hogg records that not only was it considered obscene for a man to show a woman the way to the lavatory, but that even man to man, or woman to woman, an evasive phrase had to be used: "Would you care to wash your hands?" "Have you been shown the geography of the house?"

And he makes no excuse for Victorian sexual mores or their consequences, for example, to women married in complete ignorance. Nor is he incapable of laughing at himself, or so I would interpret a passage in which he quotes a mythical future ethnologist turning up evidence of strange contemporary custom:

> . . . he clings to the very superstition which he records among primitive tribes, that to dispatch the tribal god by indirect means is not blasphemy in the first degree: that is, he treats facetiously the beliefs and ceremonies of almost every religion but that of contemporary English Protestantism, but points out the common resemblances and leaves the reader to take the inevitable step . . .

And farther along he says, in the same vein,

> While he speaks with bantering condescension of the poor savage who uses the navel-cord and severed genitals of his relatives for the magic purposes of agriculture, the language he chooses is

blamelessly scientific. In other words, he gives himself the privilege of the priests who may treat of the holy mysteries plainly, but in the sacred language and not in the vernacular . . .

Still later he calls the same speaker "exquisitely circumlocutory." The phrase, as he certainly knows, fits his own word "lavatory" (which appears twice in the first quotation on page 82).

Then, on the other hand, Graves sets a limit to his criticisms:

> To consent uncritically to the taboos, which are often grotesque, is as foolish as to reject them uncritically. The nice person is one who good-humouredly criticizes the absurdities of the taboo in good-humoured conversation with intimates; but does not find it necessary to celebrate any black masses as a proof of his emancipation from it. This book is written for the Nice People . . .

As he comes to the end of his book the author, having up to that point sustained a mood of lighthearted half-seriousness, shifts abruptly to an intensity expressive of his own deeper feelings, and it is here especially that I find myself disagreeing with him. We must have his own words at length:

> Though *Ulysses* can be studied as a complete manual of contemporary obscenity, such a study will get no encouragement here. It is true that *Ulysses* is forbidden publication in England as indecent and that it contains more words classified by law as indecent than any other work published in this century; but on the other hand it also contains more obscure poetic and religious references than any other work published in this century and the choice of language in the blameless passages is as scholarly as Mr. Saintsbury's and as English as Charles Doughty's. So far from being a work of merely pornographic intention or even a serious work given to pornographic sugar-coating that Rabelais gave his politico-philosophic pills, it is a deadly serious work in which obscenity is anatomized as it has never been anatomized before. To call Joyce obscene, is like calling the Shakespeare of the *Sonnets* lustful: true, both have had the intimate experience that their writing implies, but Joyce has brought himself as far beyond obscenity as Shakespeare got beyond that lust of which he makes frank confession . . . Joyce is read as obscene instead of successfully past obscenity: Shakespeare instead of being read as past lust is not even read as lusting.

It is not that he defends *Ulysses* and Shakespeare passionately. Especially in view of the ban on *Ulysses* at the time, the passion seems to me altogether justified. It is rather the concessions he seems to make to the enemy. For him to state that some passages in *Ulysses* are "blameless" is to grant that others are not. To argue that the book is not "*merely* pornographic" is to admit that it is partly so. Is *Ulysses* a great work because of its "obscure poetic and religious references" or precisely because in it "obscenity is anatomized as it has never been anatomized before"? Is it not true of Joyce, as it is of Shakespeare in the *Sonnets* and elsewhere, that he has given us a glimpse into the nature of man as only a great poet can show it to us? And are not the deep and usually hidden feelings, which we call obscenity in one case and lust in the other, as essential to the portrayal as any other part of it? Far from having "brought himself beyond obscenity," it seems to me, Joyce has simply refused to be limited by the prevailing prejudice. Language is the instrument of poetry, and a bowdlerized Joyce, even with its poetic and religious references intact, would still be a eunuch.

I find myself objecting to the passing poke Graves takes at Rabelais: "his politico-philosophic pills" had a "pornographic sugar-coating." Graves, I think, has granted his opponent's major premise; and, having done so, he cannot successfully refute his conclusion. Lawyers may argue the working meaning of "obscenity." They do so as advocates, with their private opinions suspended. A poet's advocacy can be only his own, and for him to concede a little obscenity is to yield the principle and lose the argument.

A few years after *Lars Porsena*, on December 6, 1933, the "monumental decision" of Judge John M. Woolsey lifted the ban on *Ulysses* in the United States. In its text, "obscene" is carefully defined on the basis of a series of legal references as "tending to stir the sex impulses or to lead to sexually impure and lustful thought"—that is, in its pornographic sense alone. The basis of the decision is that the book does not have such an effect on "the normal person." The penultimate paragraph of this famous document reads:

> I am quite aware that owing to some of its scenes "Ulysses" is a rather strong draught to ask some sensitive, though normal, person to take. But my considered opinion, after long reflection, is that whilst in many places the effect of "Ulysses" on the reader undoubtedly is somewhat emetic, nowhere does it tend to be aphrodisiac.

In a foreword to *Ulysses* by Morris Ernst, who has fought against censorship in the courts and in print for more than forty years, we are reminded that Judge Woolsey's drinking metaphor was particularly, if unintentionally, apt. Prohibition was repealed during the very week of the Woolsey decision, so that, as Ernst puts it, "we may now imbibe freely of the contents of bottles and forthright books." Perhaps it is as well that the decision on *Ulysses* was accepted without quibbling as the milestone it undoubtedly was on the road to freedom of expression. Perhaps, in fact, the good judge's distinction between "aphrodisiac" and "emetic" can be read as half a war won. If the sexually stimulating parts of a book can be resisted by "the normal person," the nauseating portions may be overlooked. But I find the whole argument specious. If one is to be titillated by books, the *National Geographic* will serve, as Lenny Bruce said. And to be nauseated by words is at best to reveal hidden emotional problems. Any suggestion that it is not the words themselves that make the gorge rise but the images they evoke must meet the undeniable truth that the same images can be evoked without penalty with less objectionable words.

The difference between the two sets of words is one of different class or station in life, and the principal objection to the words called "obscene" must therefore be a class objection. If the proscription can no longer be labeled as blasphemy or black magic, it is still, obviously or not, insubordination or rebellion. It seems to me that this is what "offense against chastity or modesty" really comes down to. Instead of being aimed at an elusive notion of decency, censorship and suppression mean in fact to put down mutiny. Chastity, after centuries of acceptance by only one sex—often hypocritical, at that—has been all but erased by resurgent feminism; and modesty, having been found inconvenient, has been swept under the rug by the great broom of advertising. Which is not to suggest that decency is meaningless or that diseased behavior is anything but deplorable and ugly in our time as in others. It is words rather than actions that concern me here, as far as they can be separated. To the degree that they cannot be, I offer the idea that the problem is one of disease and ought to be examined and treated as such. It will be remembered that the lazaretto and Bedlam taught us not to expect chains and beating to heal the sick.

It would be fatuous to deny that there are offenses against decency; they are all around us. If "obscenity" or "pornography" harms children, the effect should be capable of documentation and measurement. Could its influence be compared with that of the violence that permeates the atmosphere in which our children are growing up? Which hurts a child more, a "dirty" word he hears or sees, or the picture of handsome young adults smoking cigarettes with rapt enjoyment? Which does more damage to an impressionable young mind, a view of uncovered skin or evidence of mendacity in the politician and huckster? But even these questions seem to admit more than may be admissible. Isn't the whole traditional notion of obscenity fraudulent, or itself obscene?

Vestiges of the code of chivalry separate an aristocracy of Nice People from others and apply a notion of obscenity only to commoners. The little words properly called vulgar are not to be used publicly by Nice People, who in the same sense are not to encourage public displays of pornography. What they do in private is of course their own affair. As Mrs. Patrick Campbell said, the English don't really care what people do, so long as they don't do it in the street and frighten the horses. The great Victorian collectors of pornography were all gentlefolk, the hobby being an expensive one. (My namesake, the Fifth Earl of Rosebery, onetime Liberal Prime Minister of England, was one of them.) Their peers regarded it with good humor and indulgence; obviously it didn't hurt anybody. Yet somehow, entirely similar activity among lesser folk in the terms they understand is harmful and must not be tolerated. Even Graves seems to concede that it must be stopped, and tacitly agrees to the use of force in doing so. Accepting this idea, is he not then compelled by his own logic to object when his fellow poet Joyce seems to step across the class line and behave—in his cultivated way, of course—very much as the vulgar folk do?

A phrase that comes down to us from Elizabethan English throws a sidelight on some of these class distinctions. It is *sir-reverence*, used as a euphemism for excrement in such phrases as "to tread in sir-reverence." Bourke (whose treatise on scatology we are coming to) suggests that it was a corruption of "save reverence," itself derived from "saving your reverence." He speaks of

> an ancient custom which obliges any person easing himself near the highway or footpath, on the word "reverence" being given him by a passenger, to take off his hat with his teeth, and, without moving from his station, to throw it over his head, by which it frequently falls into the excrement. This was considered as a punishment for a breach of delicacy.

But more to the point is the suggestion that the phrase was used apologetically when mentioning anything deemed improper or un-seemly, in quite the same way that those today who are accustomed to "vulgar" speech among themselves apologize to their betters when a word slips out. Sir John Harington uses the phrase in his prefatory letter to his book, which we shall look at more closely in the next chapter. He speaks of

> the old saying, pens may blot, but they cannot blush . . . this same excellent word save-reverence, makes all mannerlie.

Sir John was of course playing at apology, as Elizabethans liked to do. The true apology to one of a class above would seem to have been based on the supposition that the air was purer at higher altitudes, and was thus an effort to dissipate assumed pollution. Bourke, still speaking of "sir-reverence," misses this point, I think, when he says:

> We can hardly credit that peasantry living in an age when the highest classes received their guests at bedside "ruelles," or in their "cabinets d'aisance" [privies], should be squeamish in the trifling matter just alluded to.

It was not that the peasantry were squeamish, but only that since they could not understand what the gentry were saying they assumed that their own language was private, too; and perhaps there was an overtone of sympathetic magic, the fear that the word might be trans-formed into the thing.

In our own time practices marked as obscene have been beaten down only to rise again. New methods, from photography to the paint-spray can of the New Graffiti, have quickly been pressed into service. If in the course of time occasional words have disappeared from the language, force from above is hardly to be credited with a kill; new ones rose to take their place, and the principal old words lived on. Today, with monosyllables referring to sex and excretion forbidden in newspapers and on television, they are nevertheless dis-played and available at low cost in the paperback book racks of drug-store and airport, and there are specialized centers for such books near Times Square in New York City and doubtless somewhere in every city in the United States. They are part of the regular fare of today's theater and are finding their way into movies labeled "for adults only" (sometimes even "for mature adults only"). Movies and still pictures in magazines, especially those accessible in price and content to the relatively uneducated poor, seem to be the principal current targets of the censors. We live, to be sure, in an age that is surely corrupt if not depraved, but causes and effects are intermingled, and actual damage done by so-called obscenity is yet to be measured.

Suggestions of a double standard in the enforcement of customs and laws against obscenity, which tend to deny certain enjoyments only to those with small purses, find a parallel in the tastes of the enforcement agents. I don't mean those at higher levels, and will grant for simplicity's sake that the self-appointed Comstocks may be unassailably pure, if desiccated. Lower down, however, the police, who sometimes seem to exercise legislative and judicial as well as executive functions at the operative level, are themselves drawn from the suffering class and usually share both its vocabulary and its enjoyment of the forbidden fruit. They exercise these rights, of course, only in the permitted environment of locker room or latrine. And so when they are witness to the punishable offenses I am speaking of, it cannot be the crime itself but only the circumstances under which it is committed that inflames in the breasts of these good people an outrage that sometimes explodes.

Consider the case of Lenny Bruce. I never saw him perform and judge him by his books and records and the reinforcing comments of a few who knew him. I find him to have been a latter-day Mark Twain, with his own special stamp. Bruce, who performed by talking from a stage, had Mark Twain's unmasked vision of "the damned human race" and the same genius for making human hypocrisy explosively funny. But he put together in his performances the language Mark Twain used in his privately circulated little book called *1601*, and the searing philosophy he displayed shyly in *The Mysterious Stranger*, but otherwise left for posthumous publication. Lenny Bruce, in fact, rushed in where Mark Twain feared to tread openly, and the consequence was to detonate the anti-obscenity machinery. Repeatedly harassed by the police, brought to court again and again for obscenity, and later on narcotics charges that may well have been trumped up, Bruce seems to have been guilty only of telling truth too directly in a forbidden manner of speech. The harassment destroyed him. Paul Krassner said he "died, an official autopsy showed, from an overdose of police." Kenneth Tynan, in words written while Bruce was still alive, applied to him Flaubert's "the artist is a disease of society" by calling him "a disease of America." Dick Schaap ends the Afterword of Bruce's autobiography with a crisp comment echoing what I have been suggesting, that obscenity lay not in the man or in his words but in the circumstances of his death at the age of forty.

9

Toilet Training

[Of latrines] their introduction cannot be ascribed to purely hygienic considerations, since many nations of comparatively high development have managed to get along without them; while on the other hand tribes in low stages of culture have resorted to them.

—J. G. Bourke, *Scatologic Rites of All Nations*

According to the British architect Lawrence Wright, in his book on the history of the flush toilet or water closet, the first patent for such a device was issued in 1775, and it was not until the bacteriologic era—between 1870 and 1900—that its use became common. No doubt dramatic experience with cholera and typhoid fever supplied the final push. The result has not been an unmixed blessing. What we have come with characteristic euphemism to call the bathroom or lavatory (in English inns the former label on a door may reveal nothing but a bathtub behind it) has achieved a place in our lives on which it would be interesting to have an ancient civilized perspective, say from Aristotle or Shakespeare. Some idea of what has happened to us may be suggested by three different exhibits: a satirical piece by anthropologist Horace Miner, a recent book simply called *The Bath Room* by architect Alexander Kira, and a current advertisement for *"Bijoux des Bains."*

Professor Miner's piece is called "Body Ritual among the Nacirema" and purports to be an account of the strange customs of a people with "a highly developed market economy" who are quickly identified by spelling the tribe name backward. They spend much of their time in economic pursuits, but a large part in ritual activity, the focus of which is the human body, the appearance and health of which loom as a dominant concern in the ethos of the people. While such a concern is certainly not unusual, its ceremonial aspects and associated philosophy are unique.

The fundamental belief underlying the whole system appears to be that the human body is ugly and that its natural tendency is to debility and disease. Incarcerated in such a body, man's only hope is to avert these characteristics through the use of the powerful influences of ritual and ceremony. Every household has one or more shrines devoted to this purpose. The more powerful individuals in the society have several shrines in their houses and, in fact, the opulence of a house is often referred to in terms of the number of such ritual centers it possesses. Most houses are of wattle and daub construction, but the shrine rooms of the more wealthy are walled with stone. Poorer families imitate the rich by applying pottery plaques to their shrine walls.

While each family has at least one such shrine, the rituals associated with it are not family ceremonies but are private and secret. The rites are normally only discussed with children, and then only during the period when they are being initiated into these mysteries . . .

Details follow of "a box or chest . . . built into the wall," in which "are kept charms and magical potions without which no native believes he could live"; of "medicine men" and "herbalists" who, having been "rewarded with substantial gifts" decide upon the contents of these potions and provide them; and of "holy-mouth-men" and the daily mouth-rite. This

involves a practice which strikes the uninitiated stranger as revolting. It was reported to me that the ritual consists of inserting a small bundle of hog hairs into the mouth, along with certain magical powders, and then moving the bundle in a highly formalized series of gestures.

Descending from Professor Miner's Olympian perspective, we find Professor Kira down-to-earth. His book is touted on its flyleaf as "THE UNIQUE, FASCINATING AND SHOCKING STUDY OF AMERICA'S 'UNMENTIONABLE' ROOM." Subtitled "Criteria for Design," it turns out to be the report of extended research at Cornell University sponsored jointly by the university's Agricultural Experiment Station and by a plumbing company. The author is an architect, and his thesis is that bathroom furniture is archaic and urgently needs to be redesigned. He details careful investigations of persons engaged in the various bathroom activities, illustrated with drawings and photographs, the latter of women wearing bathing suits. He then proceeds to furnish detailed "Design Considerations for Cleansing" and similarly for defecation and urination. Figure 34 is a diagrammatic "plan view of sitting position on a conventional water closet" showing superimposed on the outline of a toilet seat an evidently bisexual figure and locating the positions over the seat opening respectively of the anus, the ischial tuberosities (bony foundations of the buttocks), and the female and male orifices, represented appropriately by the conventional symbols for Venus and Mars. The legs and feet are shown in two positions, the smaller feet of the inner or closer position being explained thus in the text:

> In sitting normally on a standard seat the weight is also borne by the back of the thighs. Women tend to be restricted by their clothing to sitting with their legs together and thus receive support all along the backs of the thighs. Men, on the other hand, as illustrated in Figure 34, sit with their legs spread apart and concentrate their weight at about the midpoint of the thighs.

Professor Kira affirms, apparently without question, the idea that a

> notion of total cleanness [*sic*] is a concept which appears to be basic to all peoples and to be generally valued in all cultures,

although he does observe in passing that

> in some countries it is still quite socially acceptable to smell of honest sweat.

But, he says,

> as physicians and anthropologists have pointed out, our various body odors are largely a carryover from primitive man, unnecessary in the present day.

Exhibit three depicts a flat metal box with cover raised enough to let a diamond bracelet, necklace, and ring spill out on a table. Attached to the box lid is a bejeweled set of faucet and two handles. The text speaks of a lady who has lots of diamonds, which her husband calls "hardware," jewels that spend more time in a vault than on the lady's person, especially when she is at home. It suggests that she could add facets of dazzle and brilliance to her home life by means of the sparkling elegance of beautiful faucets, like these in cut crystal, richly mounted in 24K gold plate. Elegant bathroom fixtures are as enduring as diamonds, yet cost far less (these are $216).

(I remember a remark made in a lecture by Professor Joseph Mac-Farland at the University of Pennsylvania when I was a student there. In speaking of dementia paralytica, also called general paralysis of the insane, a manifestation of neurosyphilis, he detailed among the symptoms certain delusions of grandeur that may show themselves, for example, by the sudden decision of a hitherto modest man to refurbish his bathroom in onyx and gold!)

Professor Kira builds on the assumption that "total cleanness" is desirable. He argues strongly and repeatedly that our cleanliest practices are not nearly good enough. Aside from the natural difference one would expect between a bacteriologist and an architect in interpretation of the phrase "total cleanness," it must be clear already, and will become more so, that I proceed on a different assumption. We have had a close look at the word "dirt" and have found it to mean several things, not all of which are evil. Professor Kira's word "total" seems to me inappropriate whether we observe it through a microscope or with the naked eye, or even with our noses.

I know of four books that concentrate actually or symbolically on a single item of "bathroom" furniture, that miracle of modern plumbing by which civilizations are judged, the toilet bowl or water closet. One of these, which I have mentioned before, is Sir John Harington's *The Metamorphosis of Ajax,* an Elizabethan masterpiece in which is described, in an appropriate literary framework, what was probably the first proved instrument of this genus—or its reinvention (as we shall see) after a total eclipse during the Dark Ages. Two others are recent works. Reginald Reynolds' *Cleanliness and Godliness* was written in England during World War II. The author says he

> began this book to escape from a world that wearied me, and believing that the best part of man went down the drain . . .

Lawrence Wright's *Clean and Decent,* published in 1960, combines scholarship with plumbing technology. These three books are all lively and informative; but the fourth is in a class by itself.

The fourth book, which I have spoken of several times before, is the 496-page treatise *Scatologic Rites of All Nations,* compiled by Captain John G. Bourke, Third Cavalry, U.S.A., and published in 1891. On its title page appears the following descriptive matter and advice:

A Dissertation upon the Employment of Excrementitious Remedial Agents in Religion, Therapeutics, Divination, Witchcraft, Love-Philtres, etc., in all Parts of the Globe. Based upon Original Notes and Personal Observations, and upon Compilation from over One Thousand Authorities. Not for General Perusal.

As I have already mentioned, Captain Bourke begins with a series of disclaimers that had me fooled at first. He intersperses his exposition with such editorial words as "horrible," "nastiness," "unseemingly," "lewd and vulgar," "disgusting," and others that suggest aversion to the whole enterprise. Early in the book he would have us believe that he is unequivocally on the angels' side, with this apology for the Puritans:

> The Puritan's horror of heathenish rites and superstitious vestiges had for its basis something above unreasoning fanaticism; he realized, if not through learned study, by an intuition which had all the force of genius, that every unmeaning practice, every rustic observance, which could not prove its title clear to a noble genealogy was a pagan survival, which conscience required him to tear up and destroy, root and branch.
>
> The Puritan may have made himself very much of a burden and a nuisance to his neighbors before his self-imposed task was completed, yet it is worthy of remark and of praise that his mission was a most effectual one of wiping from the face of the earth innumerable vestiges of pre-Christian idolatry.

It was not until I worked my way further into this book that I took note of what looked like Bourke's cheek sticking out with his tongue behind it. As he warms to his subject he continues to throw in an occasional reminder of his piety and quotes others doing so with abandon; but most of the time he is detached in an eminently scientific way and seems to lose no chance to let versifiers and others give us the lighter side of his subject. This approach struck me as being admirable.

Scatologic Rites is not an easily accessible book, and you may be unable to find it if you look for it, unless you have a specialist's credentials. I think we can assume that its rarity is not connected with any difficulty general readers would have in understanding it—even though it is a scholarly book—or even with a particularly widespread lack of interest in it. This would seem rather to be an aspect of our immediate problem: the book deals with a subject that has been carefully buried, and so it is itself kept guarded from the common view. But the reason for doing this, according to our working hypothesis, is sympathetic magic, which we are not afraid of. So let us take a good look at Bourke.

Put together with the rest of the exhibits, the good look will occupy the rest of this chapter and two more. Bourke gives us what Frazer left out. Much of it is remarkable, and the merely remarkable I need only transmit. But some of it is astonishing, and this I have undertaken to verify. Doing so seemed the more necessary because my two more recent sources omitted this material. Wright must have overlooked Bourke. Reynolds mentions one of his papers, but *Scatologic Rites* itself seems to have been sealed away from him in the British Museum, which was closed at the time to protect its treasures against German bombs.

Our immediate topic is the history of the privy.

Bourke tells us that certain Australian aborigines, as well as natives of the Marquesas in the South Pacific, and some American Indian tribes, observed the feline custom of burying their feces or covering them with earth. Their use for this purpose of a pointed stick or other digging instrument is similar to the practice of the ancient Jews and Turks. Harington says:

> every cat gives us an example (as housewifes tell us) to cover all our filthiness, & . . . to make your selfe, your gloves, and your clothes the more sweet, refuse not to follow the example of the Cat of the house . . .

The practice of the Jews is given in Deuteronomy 23:

> 12. Thou shalt have a place also without the camp, whither thou shalt go forth abroad:
> 13. And thou shalt have a paddle upon thy weapon; and it shall be, when thou wilt ease thyself abroad, thou shalt dig therewith, and shalt turn back and cover that which cometh from thee . . .

Reynolds tells us that actual privies date as far back as Neolithic times. The remains of such structures were dug up in the stone walls at Skara Brae in the Orkney Islands, with primitive stone-lined sewers running under the huts to the beach. Similar remains have been found in Cornwall and in the Gironde in France.

The remains of privies, usually fashioned out of masonry and provided with drains, some still functional when unearthed, *some even provided with flush devices*, have been dotted throughout Bronze Age cities going back to the third or fourth millennium B.C. Among the earliest were three privies left by the Sumerians at Tell Asmar in Iraq. Six were found at the palace of Sargon, dating from about 2400 B.C. Reynolds mentions that these had high seats, whereas those of the Orient were at floor level. (One assumes that the ancient Chinese, whose technology was well developed, had privies, but I have seen no description of them.) In the remains of the well-developed city of Mohenjo Daro in northwest India, which also dates from the third millennium B.C., were found bathrooms with masonry privies and drains made of flanged earthenware pipe. Other privies, sewers, and drains have been found in remains of ancient Egypt. A bathroom and lavatory and a well-preserved water closet were found at Tell el Amarna in Egypt, and also at Kahun, dating from the Second Dynasty (3400–2980 B.C.). Most of these were for nobles and priests, but some of those at Mohenjo Daro were in the poorest section, in the streets as well as the houses. The palace at Knossos in Crete contained very sophisticated water pipes, 2.5 feet long, nearly 1 inch thick, and slightly tapering toward one end. This superlative Minoan structure, dating back to about 1800 B.C., boasted bathrooms and a flushing water closet with a wooden seat.

The Greeks had hot and cold showers and tubs (we know the latter from Diogenes and Archimedes) as well as privies, sewers, and aqueducts. Such things seemed to reach a peak and then decline among the later Romans, presumably with the advance of Christianity, and in the process the flush toilet evidently disappeared completely. But meanwhile the Romans built flushing latrines in Britain, their farthest outpost. Rome itself had nearly 1000 public baths, water being conveyed in lead pipes. The public latrines were social and convivial. Part of the reason for this custom was the conservation of the products for practical purposes, a separate matter to be considered later.

The Romans would have horrified Professor Kira, as they did Torquemada, whom Bourke calls "a Spanish author of high repute" and quotes regarding Romans and Egyptians as follows:

I assert that they used *to adore* . . . stinking and filthy privies and water-closets; and, what is viler and yet more abominable, what is an occasion for our tears and not to be borne with so much as mentioned by name, they adored the noise and wind of the stomach when it expels from itself any cold or flatulence; and other things of the same kind which . . . it would be a shame to name or describe.

Details of medieval and later history are contributed by Wright. The medieval word for privy was *garderobe*—a place to hang one's clothing—indicating that euphemism was already at work. In a plan of Southwall palace the privies radiated around a central shaft, facing outward onto a circular passage. Neighbors could converse with one another while decently out of sight. Where no stream or moat ran below, there might be a removable barrel, or a pit like one at Everswell in 1239:

> What the cleaning of these pits meant is shown by an account of the work at Newgate Jail in 1281, when thirteen men took five nights to clear the "cloaca" . . . the men were paid three times the normal rate . . .

After the Black Death, which was attributed to everything imaginable except the rats and fleas that actually carried it, attempts were made everywhere to keep privies clean. About 1375 the cleaners were called *gongfermors*, who

> seem to have become adapted to their horrid task, for an ancient story tells of one who complained, out of working hours, about the smell of a badly snuffed candle.

A parallel anecdote is quoted by Bourke with the date 1658:

> A certain countryman at Antwerp . . . when he came into a shop of sweet smells, . . . began to faint, but one presently clapt some fresh smoking horse-dung under his nose and fetched him to again.

We are told by Wright that Leonardo da Vinci, in his proposal for Ten New Towns, aimed to

> "distribute the masses of humanity, who live crowded together like herds of goats, filling their air with stench and spreading the seeds of plague and death." In these towns, the drainage of all private and public privies, and all garbage and street sweepings, were to be carried to the river by sewers (*via sotterane*). All the stairways in the tenement buildings were to be spiral, to prevent the insanitary use of stair landings. Leonardo invented a folding closet seat that "must turn round like the little window in monasteries, being brought back to its original position by a counterweight." . . . For Frances I at Ambrose Castle he proposed to install a number of water closets with flushing channels inside

the walls, and ventilating shafts reaching up to the roof; and as people were apt to leave doors open, counterweights were to be fitted to close them automatically.

But like Leonardo's other inventions this was never actually built.

The Christians undermined hygiene, Reynolds complains. Sanitation was alien to their architecture, and the odor of sanctity was not always that of soap. He calls a chapter "The Ordure of Chivalry." The garde-robes of the Tower of London, he says, were crude and too close to the banqueting hall. Latrines were so built as to permit of no ventilation. Even among the wealthy, the bottom layer of rushes on the floor "is undisturbed sometimes for twenty years, harbouring abominations not to be mentioned."

Harington provides testimony on this subject:

> I found not only in mine own poor confused cottage, but even in the goodliest and stateliest pallaces of this realme . . . this same whorson sawcie stinke . . . great and well contrived houses . . . have vaults and secret passages . . . under ground, to convey away both the ordure and other noisome things . . . vents . . . drawing up the aire as a chimney doth smoke. By which it comes to passe manie times (specially if the wind stand at the mouth of the vaults) that what with fishwater coming from the kitchens, bloud and garbage of foul, washing of dishes, and the excrements of other houses joyned together, and all these in moyst weather stirred a little with some small streams of raine water. For as the proverbe is,
>
> > "Tis noted as the nature of a sinke,
> > Ever the more tis stird, the more to stink."

Before giving the details of his own invention, Sir John describes two earlier devices. First,

> A close vault in the ground, widest in the bottome, and narrower upward, and to floor the same with hot lime and tarris [a rock used for mortar or cement] or some such dry paving as may keep out water and aire also . . .

He suggests that this device works so long as it is airless like a candle snuffer, hence unstirred; but "a crannie in the wall as big as a straw" spoils it. The second method is

> either upon close or open vaults, so to place the sieges or seats as behind them rise tunnes [pipes, conduits, or chimney pots] of chimneys, to draw all the ill aires upwards . . .

This was the method used at Lincoln's Inn, and helped the neighborhood but not the privies themselves, especially when the wind was wrong.

Harington's water closet, which he diagrams and presents in full working detail, seems to have been duplicated, except for the one in his own "cottage," only at the Queen's palace at Richmond. It waited nearly two hundred years to be reinvented, and another century for general acceptance. That even the rich did not see the need for such a contrivance is shown by the practices that continued in the meantime.

By the early sixteenth century the "close stool" had begun to displace the garderobe, being, as Wright puts it, "cosier for the user but harder on the servants." Harington mentions an inventory item, "ffyve close stooles of black velvet quilted, with panns."

The enclosed chamber pot has generated a considerable literature, especially the practice that seems to have been universal at the time— what else was there to do?—of throwing its contents into the street at night through an upstairs window. Hogarth illustrates the procedure in his "Night," an engraving from the series *Four Times of the Day* (1738). Dryden, translating Juvenal, is evidently speaking of his own seventeenth-century London rather than of second-century Rome when he writes:

> Return we to the dangers of the night;
> And, first, behold our houses' dreadful height:
> From whence come broken potsherds tumbling down;
> And leaky ware, from garret windows thrown:
> Well may they break our heads, and mark the flinty stone.
> 'Tis want of sense to sup abroad too late;
> Unless thou first hast settled thy estate.
> As many fates attend thy steps to meet,
> As there are waking windows in the street.
> Bless the good gods, and think thy chance is rare
> To have a pisspot only for thy share.

Bourke says there were no privies in Madrid in 1769, and the attempt by the king to introduce them, as well as sewers, "and to prohibit the throwing of human ordure out of windows after nightfall, as had been the custom, nearly precipitated a revolution." But in Paris there were royal ordinances as early as 1372 and again in 1395—doubtless inspired by the Black Death—against throwing ordure out of windows.

By 1738 we find Blondel, in France, speaking of *cabinets d'aisance à l'anglaise* or *lieux à soupape* (valve closets) as having originated in England; but they did not exist in London at the time. Wright mentions that at Windsor, Queen Anne had "a little place of Easement of marble with sluices of water to wash all down"; while a water closet reminiscent of Harington's is described in 1718 at Sir Francis Crew's house in Beddington in Surrey. But even after 1775, when Alexander Cummings took out a patent for a device with all the elements of a modern water closet, its acceptance came only slowly. One of the possible reasons is suggested by the drawing by Rowlandson, circa 1790, entitled "Work for the Plumber." It may have been harder to cope with bad plumbing than with none.

As late as 1858 "public conveniences" were still scarce and foul in London. Wright gives credit for winning the battle for better ones to George Jennings, about 1870; but the stranger abroad in an American city today may have legitimate grounds for complaint on this score. The modern flush toilet dates from about 1908, with improvements about 1935. But the outhouse and its Sears-Roebuck catalogue have not yet become obsolete.

There are reasons for believing that an accessory to the latrine has always been required by defecating human beings, although animals seem to get along quite well without it. The same accessory is less universally applied after urinating. Kira has words of glowing tribute for the bidet, which he thinks has acquired an evil reputation by association with an alleged sexual specialization. If not for this, he suggests, it might serve admirably toward his goal of total cleanness. In the absence of the bidet, we take for granted the use of paper. Men customarily limit its application to the anus. Kira recommends that they would do well to follow women in using it after urinating as well.

Means toward this end (indeed toward both ends) were not lacking before paper was available. The following passage is attributed by Bourke to Arminius Vambery (London, 1868):

> The manner of cleaning the body after an evacuation of any kind is defined by [Mohammedan] religious ritual. "The law commands 'Istinjah' (removal), 'Istinkah' (ablution), and 'Istibra' (drying)"—i.e., a small clod of earth is first used for the local cleansing, then water at least twice, and finally a piece of linen a yard in length. . . . In Turkey, Arabia, and Persia all are necessary, and pious men carry several clods of earth for the purpose in their turbans. "These acts of purification are also carried on quite

publicly in the bazaars, from a desire to make a parade of their consistent piety." Vambery saw "a teacher give to his pupils, boys and girls, instruction in the handling of a clod of earth, and so forth, by way of experiment."

An earlier author, Tournefort (1718), is quoted as saying:

> Moslems urinate sitting down on the heels; for a spray of urine would make hair and clothes ceremonially impure. . . . After urinating, the Moslem wipes the os penis with one to three bits of stone, clay, or a handful of earth.

Mohammedans were evidently known also to drain the last drop of urine by touching the penis to a stone wall. Bourke mentions a practical joke played on them by Christians who put hot pepper on the wall. One assumes that the practice was looked upon as one of the pagan survivals which conscience required the Puritans "to tear up and destroy, root and branch."

Bourke cites Pinkerton (1814) as saying that, among the Chinese,

> it is usual for the princes, and even the people, to make water standing. Persons of dignity, as well as the vice-kings, and the principal officers, have gilded canes, a cubit long, which are bored through, and these they use as often as they make water, standing upright all the time; and by this means the tube carries the water to a good distance from them.

Bourke tells us that in Rome a bucket filled with salt water was placed in each public latrine, with a stick in it having a sponge tied to one end. With this instrument "the passer by cleansed his person, and then replaced the stick in the tub." Reynolds also speaks of this procedure and provides further details. After mentioning in passing *mempiria*, or balls of hay "which our medieval ancestors used [for] the final polish," he describes a medieval tool shaped like a hockey stick, and gives it as the source of the old phrase "to get hold of the wrong end of the stick." But in extenuation he emphasizes that "our horror of excrement . . . is really a very newfangled idea." He cites a story told by Sir Roger L'Estrange of King James I,

> that such was his absorption in this sport [hunting] that he would not leave the saddle even to relieve himself, so that his servants had a pretty mess to clean up at the day's end.

And since Sir Roger was an ardent royalist the story was not told to discredit the monarch.

Coming back to paper, Reynolds said of the wartime shortage:

> the day may soon come when we shall have to choose between
> a pound note and a petrol coupon, unless we go back to the
> Roman sponges or . . . mempiria.

He mentions an unnamed person who wrote to his enemy, "I have
your letter before me, and it will soon be behind me"; and goes on
to recommend a paper

> sufficiently thick but sufficiently pliable, glazed upon one side
> only and bearing, perhaps, some cheerful distich . . . to elevate
> the mind rather than to advertise the product or to proclaim each
> sheet to be the property of the Government or the Corporation
> of Margate.

As recently as 1962, when I was in London, I did indeed find the
product in a public convenience marked, on every sheet, "Property
of the British Empire." My good friend Dr. Max Pleasure, a dentist
with a wit to match his name, now dead and much lamented, once
wrote me from Israel, enclosing a sample of the local material, a very
poor product indeed by lavish American standards, with the glaze on
one side that Reynolds must accept as an irreducible minimum. Max
had written on the glaze, his dental technology showing, "This side
for polishing."

Nobody seems to have collected the international practices of today.
Another old friend of mine could have done a good job of it if he
had lived. Dr. Leon Buchbinder was an eminent bacteriologist and
a longtime sufferer from ulcerative colitis. He became a collector of
latrines, and in New York, and doubtless in many other places, he
always knew the most direct path to a nearby one that met his standards. My own experience abroad is modest. At a *pension* on the Left
Bank in Paris one private "bathroom" contained a bidet and a lavatory,
and another provided a bathtub; the toilet was in the public hall,
halfway between floors. In Moscow at the Hotel Ukraine the fixtures
seemed to me somewhat overdone in splendor, especially the porcelain-
and-nickel fittings over the tub for both regular and hand showers.
Everything worked adequately but there was no shower curtain or
other means of limiting the flood. By contrast the Hotel Malmen in
Stockholm was elegant in its modernist simplicity; but a fixed glass
panel in front of the shower restrained only a portion of the flood,
and the toilet bowl dripped all night.

We think of our water-closet practices as universally, or at least as uniquely, civilized and correct. But customs vary surprisingly. Bourke tells us that Apache men in urinating always squat down, while women stand up; and similar reversals of our practice have been recorded in many other places. Montaigne mentions "people . . . where women piss standing, and men cowering." Dryden gives us these lines in his translation of Juvenal's sixth satire:

> Behold the strutting Amazonian whore!
> She stands in guard, with her right foot before;
> Her coat tucked up, and all her motions just,
> She stamps, and then cries, "Hah!" at every thrust.
> But laugh to see her, tired from many a bout,
> Call for the pot, and like a man piss out.

This posture seems to me strikingly similar to that shown widely today in women's magazines and elsewhere in advertisements for miniskirts.

Reynolds has compassion for male visitors in London accustomed to squatting and faced—or backed—in public conveniences with nothing but a vertical target.

Professor Miner's rule of sacred privacy is typically violated today in the lower ranks of the military and without duress by physicians. But at one time there was a grace in such matters that we have lost. It seems to have begun in Rome but came to be called "the French courtesy." Harington tells us of it:

> I have heard it serious told, that a great Magnifico of Venice, being ambassador in France, and hearing a Noble person was come to speak with him, made him stay till he had untyed his points; and when he was new set on his stoole, sent for the Noble man to come to him at that time; as a verie speciall favour.

And again, Sir John says of Martial that

> in his third booke, he mocks one of his fellow Poets, that drave away all good companie with his verses, everie man thought it such a penance to heare them . . .
> Alas my head with thy long readings akes,
> Standing or sitting, thou readst every wheare,
> If I would walke, if I would go t' A'JAX,
> If to the Bath, thou art still in mine eare.
> Where by the way, you may note, that the French courtesie . . . came first from the Romanes; sith in Martial's time, they shunned not one the others companie, at Monsieur A JAX.

(Harington has explained that the name of the burly and somewhat dull-witted Greek warrior came to be applied to the privy, Ajax being sometimes written A JAX, and then changed further to "a Jakes." This, and the genealogy that heads the next chapter, is "the metamorphosis of Ajax.")

Reynolds regards reading on the toilet as a masculine habit. "This I am inclined to believe," he says, "from my experiments in timing the two sexes." He holds that Rodin's *Le Penseur* is "utterly unimaginable" in a woman. Associated with this sex difference may be the greater propensity of men than of women to inscribe graffiti on latrine walls. Bourke makes only passing mention of this art form, citing as reference without examples an anonymous and otherwise unidentified "Bibliotheca Scatologica," of which neither Reynolds nor I could find any other sign. Reynolds himself records notable examples of graffiti, and a wider range has been collected and published recently, uninhibited, and with an absurd suggestion of copyright. Reynolds assures us that it is for the

> unhallowed pleasure of breaking *taboo* that generations of writers and artists have produced upon lavatory walls since the days of Herculaneum and Pompeii (and no doubt at Knossos and Mohenjo-Daro, if the whole truth were known) those curious inscriptions and sketches with which every age has become familiar . . .

His examples are scholarly, or anyway hardly to be thought of as vulgar in terms of origin. From Merton College:

> ; everything passes, nothing remains.
> —Aristagoras

Beneath which, in another hand:

> ; everything remains, nothing passes.
> —Constipagoras

Said to have been inscribed either at Somerset House or at the Inns of Court:

> I do not like this place at all,
> The seat is too high and the hole is too small.

And again, a rejoinder below by another visitor:

> You lay yourself open to the obvious retort
> Your bottom's too big and your legs are too short.

More typical of a group famliar to collectors is this one in the "convenience" below the Halls of Parliament:

> In the House above, when a motion is read,
> The Member stands up and uncovers his head;
> In this House down below when a motion's to pass,
> The Member sits down . . . [*sic*]

The physician Oliver St. John Gogarty, who was the prototype of Joyce's Buck Mulligan, has given us the following distinctively Irish story in his own book:

> . . . we met Thwackhurst, the collector of graffiti. . . . "God blast you and them, anyway," said Thwackhurst cheerily. "You have ruined all the graffiti. You can't find anything in a urinal now but political remarks. It's always a sign of decline in the fortunes of a country. Even at their moments of ease the people are obsessed with thoughts of politics. Instead of thinking of the matter in hand. And just when the spread of popular education was bringing the graffiti lower on the walls."
> "Lower on the walls?"
> "Sure. Don't you see the little children were beginning to add their quota, when all this damn politics comes along."
> "What would you consider as a graffito of the Golden Age?"
> Thwackhurst fumbled for a note-book, but before he found it he started reciting:
> > "Here lies the grave of Keelin,
> >   And on it his wife is kneeling;
> > If he were alive, she would be lying,
> >   And he would be kneeling."
> . . . "God blast you and your politics," Thwackhurst repeated. "They are as bad as glazed tiles! You have ruined the finest town for graffiti in Europe, and ended its Elizabethan age."

As an ardent champion of organic fertilization of soil, Reynolds, in attributing the following epitaph to Byron, suggests that it is suitable for all statesmen:

> Posterity will ne'er survey
> A nobler grave than this;
> Here lie the bones of Castlereagh.
> Stop, traveller . . . [*sic*].

Let me end this chapter with a few miscellaneous items, all from Bourke. He cites from a fourteenth-century farce:

A miller, believing that at the moment of death his soul escapes through the anus, arranged that his wife and the priest pull him to the ends of the bed so as to witness the event. "The phenomenon of rectal flatulence is now observed, when suddenly, to the consternation of the wife and priest, a demon appears and placing a sack over the dying man's anus, catches the rectal gas and flies off in sulphurous vapor."

A suggestion that originally the "blood" in

> Fee, fie, fo, fum,
> I smell the blood of an Englishman

may have been something else is found in an analogous myth among the Apaches, in which the giant followed a stronger scent.

And a succession of merry verses of obscure origin, all bearing on the unexceptionable premise that water quitting the body through one orifice leaves that much less to quit it through another:

> If love demands weeping, oh, why should I spare
> Those floods which, of course, must be lavished elsewhere?

> And midst their bawling and their hissing,
> They cried, to keep themselves from p————g [*sic*].
> Finding their water would come out,
> They thought it best, without dispute,
> Rather than wet both breeks and thighs,
> To let it bubble through their eyes.

> What if she whine, shed tears, and frown?
> Laugh at her folly, she'll have done;
> Never dry up her tears with kisses,
> The more she cries, the less she p————s [*sic*].

And finally the following, dealing with the effect upon a lady of witnessing the tragedy of *Cato*, is the converse of the preceding and seems to reflect on the futility of self-control, or perhaps on the incompressibility of water:

> Whilst maudlin chiefs deplore their Cato's fate,
> Still, with dry eyes, the Tory Celia sate;
> But though her pride forbade her eyes to flow,
> The gushing waters found a vent below.
> Tho'n secret, yet with copious streams she mourns,

Like twenty river-gods, with all their urns.
Let others screw on hypocritic face,
She shows her grief in a sincerer place;
Here nature reigns, and passion, void of art,
For this road leads directly to the heart.

10

The Romans Had a God for It

AJAX sonne of Telamon,
  sonne of Aeacus.
  sonne of Juppiter.
Juppiter, *alias dictus* Picus.
  sonne of olde Saturne.
Alias dictus Stercutius.

  —Sir John Harington

The aversion we feel to man's parts and functions, and especially to his principal excretory products, is something relatively new. Primitive and ancient man felt differently about them. He accepted them, where we deny them. Vestiges of his acceptance have not entirely disappeared even now. This thesis underlies Bourke's *Scatologic Rites*. He provides overwhelming evidence of the process of change. Yet he writes as one converted to the idea that what has happened is good, as though he wishes us to believe that early notions and feelings were heathenish, barbarous horrors. At the same time, he could not really have believed this altogether, since he evidently enjoys and shares with us the pleasures that progress would deny us.

If we can project our thoughts back to pre-Christian and perhaps pre-Hebrew times, we may recognize two sets of views of these matters. One was based only on feeling; the other was intermixed with rationality. The first was a manifestation of belief in sympathetic magic. It is the counterpart in the primitive adult of the approach to his own body that we see in infants today. Notions of aversion or rejection are entirely absent, but strong feeling is present, which tends to generate fantasy. What belongs to us does not stop being ours when it is separated from us; as the embodiment of self it continues with a fantastic existence which somehow sustains its essential link to self.

The other set of views applied especially to the more reactive and changeable excretory products, which happen also to be the most abundant: that is, to urine and feces especially, in contrast to hair and nails. The latter are very insoluble and durable. They do not entirely resist the action of microbes (certain fungi destroy them), but they can last a long time. Early man would have tended to associate with them properties based not on what they can do but on what they are and remain. But feces and urine change rapidly, as both eyes and nose cannot avoid knowing; and as they do so they show diverse properties which early man, untrammeled by aversion or abnegation, must have been quick to appreciate and make use of. Some of these properties doubtless seemed visibly and tangibly—and not just symbolically—magical. Hence his early notions and feelings about them would necessarily have gone beyond sympathetic magic.

It is certainly significant that in such widely scattered developing civilizations as the Mexican and those of our Mediterranean ancestors, early notions about supernatural qualities of feces and urine crystallized in deification. Bourke tells us that the Mexican goddess *Suchiquecal*, the mother of the human race, was represented in a state of humiliation as eating *cuitlatl*, which is identified as the Greek *kopros* or Latin *faeces*:

> The vessel in the left hand of Suchiquecal contains *"mierda"* (Span.). The Mexicans also had a "Goddess of ordure," or Tlaçolquani (also given as Tlaçolteotl or Ixcuina), the *eater of ordure*, because she presided over loves and carnal pleasures. . . . Mendieta mentions her as masculine, and in these terms: The god of vices and dirtiness, whom they called Tlazulteotl.
>
> Bancroft speaks of "the Mexican goddess of carnal love, called Tlazeoltecotl, Ixcuina, Tlaçolquani," etc., and said that she "had in her service a crowd of dwarfs, buffoons, and hunchbacks, who diverted her with songs and dances and acted as messengers to such gods as she took a fancy to. The last name of this goddess means 'eater of filthy things,' referring, it is said, to her function

of hearing and pardoning the confessions of men and women guilty of unclean and carnal crimes."

. . . This goddess and another, Ochpanistli, are called by Spanish monks "La diosa de basura o pecado."

Bourke points out in a footnote that *basura* is defined merely as "rubbish" or "refuse"; but like the word *filth* it may also mean "dung, manure, ordure, excrement." The whole passage is not to be taken too literally, since its context warns us that what is really early Mexican in it is obscured somewhat by the feelings and prejudices of the Spanish monks and others from whom the information comes. The quoted words "dirtiness," "filthy things," "unclean," and even *basura* and *pecado* (sin) themselves may have been monkish interpretations of words with different original meanings. Were the interpreters apologizing for the Mexican deities, or did the old Mexicans intend to do so themselves?

Closely parallel with these Mexican deities were the Baal gods of the Canaanite and Moabite polytheists, especially Baal-Peor or Bel-Phegor. These were again associated with orgiastic fertility cults, that is, they were dung-gods believed to provide abundance of crops. Again they have come down to us in the monotheistic Hebrew Bible with a cover of pious aversion. The following quote is given by Bourke from Dulaure (1825), who cites earlier authors

> to show that the Israelites and Moabites had the same ridiculous and disgusting ceremonial in their worship of Bel-phegor. The devotee presented his naked posterior before the altar and relieved his entrails, making an offering to the idol of the foul emanations.

Deuteronomy 29:17 reads:

> And ye have seen their abominations, and their idols, wood and stone, silver and gold, which were among them:

Bourke cites interpretations suggesting that "abominations" be read "dung gods" and "idols" be read "detestable things." A further suggestion is given in a cited footnote:

> Dungy gods from the shape of the ordure, literally thin clods or balls, or that which can be rolled about.

Further along Bourke offers another suggestion, admitted in the context to be disputed, which I have translated from the French as Bourke gives it for reasons unexplained:

Certain antiquarians thought they could identify the Roman god Crepitus [of whom more shortly] with Bel-Phegor, or Baal-Peor, of the Syrians, Phegor, we are told, having this sense in Hebrew.

Whether—and I am inclined to doubt it—the original worshipers of these gods felt anything like shame in connection with them, it is perfectly clear that the Romans felt nothing of the sort. They had a group of gods concerned with the complex of excrement and fertility, and invested them with dignity. The first of these was Stercutius (*stercus* is another Latin synonym for dung), whose origin Harington gives as follows:

> *Stercus* was *Saturnes* father, hee that invented manuring of the ground with dung, which of him was called *Stercus*: Some say they called him *Stercutius*. Well howsoever he gotte the name of *Saturne*, he was the same *Stercus* or *Stercutius* whom they deified for his husbandry.

Bourke, in a passage which he also sees fit to give in French, and in which Pliny and Persius as well as Augustine are cited as sources, mentions several "Stercus-gods," including Stercus or Sterces, inventor of manuring and father of Picus, and Stercutius, alias Sterculius, Sterquilinus, or Sterquiline, thought to be names given to Saturn as inventor of agriculture. In the quotation from Harington at the head of this chapter, Ajax, whose name came to be applied to the privy, is set forth as the lineal descendant of Saturn, also called Stercutius. Saturn was the father of Jupiter, who, Ovid said, was "father of us all." The child's identification of himself with feces, as Freud has conceived it, would seem to have very deep roots in human history.

Second in importance to Stercutius among Roman deities of this group is Cloacina, who

> was one of the first of the Roman deities, and is believed to have been named by Romulus himself. Under her charge were the various sewers, privies, etc., of the Eternal City.

A footnote here reads:

> There is another opinion concerning Cloacina—that she was one of the names given to a statue of Venus found in the Cloaca Maxima.

Cloaca is the Latin word for "sewer"; it also means "privy" and is the zoological term for the common or associated genital and excretory orifices of many animals. The association of Cloacina with Venus is another early example of what Freud identified in the young as cloacal confusion. We may infer that discriminating anatomy of the human female developed slowly in the race as it does in the individual. Biologists may recognize a parallel with the principle of animal development: "Ontogeny recapitulates phylogeny."

Reynolds offers the following graffito, whose "stately lines" he attributes to Byron:

> O Cloacina, Goddess of this place,
> Look on thy suppliants with a smiling face,
> Soft, yet cohesive let their offerings flow,
> Not rashly swift nor insolently slow.

Another Roman deity, Crepitus (Latin, a crackling sound, flatulence), is traced by Bourke somewhat obscurely (in French) to Egyptian roots, and is also associated with Bel-Phegor. "Le Pet [fart] was an ancient Egyptian goddess; she was the personification of a natural function . . ."; and from Minucius Felix, in the second century A.D. (my translation):

> Crepitus . . . was perhaps only a caricature conceived by the jesters of that time. Menage [an authority?] however affirms that the Pelusians [a people of Lower Egypt] worshipped *le Pet.*

This is followed by the passage given earlier, suggesting the identity of Crepitus with Bel-Phegor. Bourke's quotation from Torquemada given in the previous chapter was clearly based on the same idea. In another place Bourke quotes from Charles Percy, M.D., who wrote in *A View of the Levant* (1743):

> The ancient Pelusians . . . did (amongst other whimsical, chimerical objects of veneration and worship) venerate a Fart, which they worshipped under the symbol of a swelled paunch.

Harington translates this verse from the Latin of Thomas More:

> To breake a little wind, sometime ones life doth save,
> For want of vent behind, some folk their ruine have:
> A powre it hath therefore, of life, and death expresse:
> A king can cause no more, a cracke doth do no lesse.

The word *cracke* is defined in a footnote as *ventris crepitus.*

Another reference in Bourke to a deity of this sort is attributed by him to Andrew Lang (*Myth, Ritual, and Religion,* 1887): a potent deity named "Pund-jel" among some of the Australian tribes, who may be the Eagle-hawk. "As a punisher of wicked people, Pundjel was once moved to drown the world, and this he did by a flood which he produced (as Dr. Brown says of another affair) by a familiar Gulliverian application of hydraulics."

Saturn having been the ruler of the world in the Golden Age of peace and plenty, his festival, the Saturnalia, celebrated for several days in mid-December, was a time to stop work and exchange gifts; and slaves were free to do as they pleased. This observance seems to have been the forerunner of Christmas, with an intermediate stage in which it became the Feast of Fools, although it appears that this celebration had other origins as well:

> The use of human and animal egestae in religious ceremonials was common all over the world, antedating the Roman Saturnalia, or at least totally unconnected with it . . . a reversion to a pre-Christian type of thought dating back to the earliest appearance of the Aryan race in Europe.

And this passage follows immediately:

> The introduction of the Christian religion was accompanied by many compromises; whenever it was opposed by too great odds, in point of numbers, it permitted the retention of practices repugnant to its teachings . . . [or] acquiesced in them.

The following then appears, with a quotation from the Reverend Fosbroke (1843):

> This ritual was written out in 1369 at Viviers in France . . . "In the Feast of Fools they put on masks, took the dress, etc., of women, danced and sang in the choir, ate fat cakes upon the horn of the altar, where the celebrating priest played at dice, put stinking stuff from the leather of old shoes in the censer, jumped about the church, with the addition of obscene jests, songs, and unseemly attitudes. Another part of this indecorous buffoonery was shaving the precentor of fools upon a stage, erected before the church, in the presence of the people; and during the operation he amused them with lewd and vulgar discourses and gestures. They also had carts full of ordure which they threw occasionally upon the populace. This exhibition was always in Christmas time or near it, but was not confined to a particular day."

A contemporary verse describes some of these high jinks:

> But others bear a torde, that on a cushion soft they lay;
> And one there is that with a flap doth keep the flies away:
> I would there might another be, an officer of those,
> Whose room might serve to take away the scent from every nose.

Bourke says this of the Druids, from whom some of these customs came:

> These sports were calculated to expose [them] to scorn and derision. The Feast of Fools had its desired effect, and contributed perhaps more to the extermination of these heathens than all the collateral aids of fire and sword, neither of which were spared in the persecution of them.

A curious derivation from the Druids dealing with the origin of the name and the use of *mistletoe* is mentioned in part by Bourke and further explained by Reynolds. Overtones of sympathetic magic are audible in this old English practice:

> The method of divination by which maidens strove to rekindle the expiring flames of affection in the hearts of husbands and lovers by making cake from dough kneaded on the woman's posterior . . . seems to have held on in England as a game among little girls, in which one lies down on the floor, on her back, rolling backwards and forwards, and repeating the following lines:
> > "Cockledy bread, mistley cake,
> > When you do that for our sake."
> While one of the party so lay down the rest of the party sat round; they lay down and rolled in this manner by turns.

This is followed by:

> Cockle Bread. This singular game is thus described by Aubray and Kennett: "Young wenches have a wanton sport which they call 'moulding of cockle bread,' viz.: they get upon a table-board, and then gather up their knees as high as they can, and then they wobble to and fro, as if they were kneading of dough, and say these words:
> > 'My dame is sick, and gone to bed,
> > And I'll go mould my cockle bread,
> > Up with my heels, and down with my head!—
> > And this is the way we mould cockle bread'."

Bourke was unable to trace "cockledy" and "mistley," and a few pages later speaks in a separate context of rue and mistletoe. Both of them had a direct, irritant action upon the genito-urinary organs, and in all probability [were] employed to induce the sacred urination and to asperse the congregation with the fluid for which holy water was afterward substituted.

Reynolds, speaking of the Druids, has this to say:

> it is a curious thing that their contribution to this festival [Christmas] should be the *mistletoe*, known to us by a name which is surely derived from *mist*, which in German signifies *dung*, and was believed by our forefathers to grow where a certain bird had left its excrement upon the branch. This *mistletoe*, with so strange an aura of dung, fertility, and love . . .

Against this interpretation, or at least not explicitly supporting it, we must set the extensive treatment of mistletoe by Frazer—mistletoe is the "Golden Bough" of his title—as well as by Bourke. Frazer deals with mistletoe in two of his closing chapters, and Bourke gives a chapter to it with a final section subtitled "The Linguistics of the Mistletoe"; but in none of this is there a suggestion of a derivation of the word from *mist*, nor, indeed, although mistletoe has been associated with various attributes—magical, medical, aphrodisiac—is there anything scatological; and its very presence in Bourke remains an enigma.

Bourke makes a few other scattered and somewhat veiled references to the use of urine as holy water, perhaps only among pre-Christian peoples. He cites Mungo Park (1813) as having observed or heard that among both African Hottentots and Moors urine was holy water, the urine of the new bride being sprinkled on the couple by the priest. In another reference to Kolbein (or Kolben), dated 1777, he says that the Hottentot priest

> "coming up to the bridegroom, pisses a little upon him. The bridegroom receiving the stream with eagerness rubs it over his body. . . ." The same is then done by the bride.

Something suspiciously similar appears toward the end of "Oberon's Feast," by Robert Herrick, in his book *Hesperides*. Oberon consumes, among other delicacies,

> The broke-heart of a Nightingale
> Ore-come in musicke; with a wine,
> Ne're ravisht from the flattering Vine,
> But gently prest from the soft side
> Of the most sweet and dainty Bride,

> Brought in a dainty daizie, which
> He fully quaffs up to bewitch
> His blood to height; . . .

A footnote to "Bride" in the fifth line of this quotation, in the version
from which I have taken it, reads: "bridewort or meadowsweet?"—im-
plying an interpretation that would not have interested Bourke, who
gives this piece. Nevertheless—and it may be coincidence, of course!
—Herrick, in "The Argument" which opens *Hesperides*, says:

> I sing of *Brooks,* of *Blossomes, Birds,* and *Bowers*:
>
>   ❋   ❋   ❋
>
> Of *Bridegrooms, Brides,* and of their *Bridall-cakes.*

Bourke gives a good deal of space to the Hindu veneration of the
cow as it extends to include urine and dung. He cites the Abbé Dubois
(1817) as having written that

> a Hindu penitent "must drink the *panchakaryam*—a word which
> literally means the five things, namely milk, butter, curd, dung,
> and urine, all mixed together." And he adds:—
>
> "The urine of the cow is held to be the most efficacious of any
> for purifying all imaginable uncleanness. I have often seen the
> superstitious Hindu accompanying these animals when in the
> pasture, and watching the moment for receiving the urine as it fell,
> in vessels which he had brought for the purpose, to carry it home
> in a fresh state; or, catching it in the hollow of his hand, to
> bedew his face and all his body. When so used it removes all
> external impurity, and when taken internally, which is very com-
> mon, it cleanses all within.

Another reference appears in Bourke which includes mention of the
urine of cows as holy water:

> The moon, as Aurora, yields ambrosia. It is considered to be a
> cow; the urine of this cow is ambrosia or holy water; he who
> drinks this water purifies himself, as the ambrosia which rains
> from the lunar ray purifies and makes clear the path of the sky,
> which the shadows of night darken and contaminate.
>
> The same virtue is attributed, moreover, to cow's dung, a con-
> ception also derived from the cow, and given to the moon as well
> as to the morning aurora. These two cows are considered as mak-
> ing the earth fruitful by means of their ambrosial excrements;
> these excrements being also luminous, both those of the moon and
> those of the aurora are considered as purifiers.

And from *Pinkerton's Voyages* (1814):

> Firewood at Seringapatam is a dear article, and the fuel most commonly used is cow-dung made up into cakes. This . . . is much used in every part of India, especially by men of rank; as, from the veneration paid the cow, it is considered as by far the most pure substance that can be employed . . . women . . . gather up the dung and carry it home in baskets. They then form it into cakes, about half an inch thick, and nine inches in diameter, and stick them on the walls to dry. So different indeed are Hindu notions of cleanliness from ours that the walls of their best houses are frequently bedaubed with these cakes; and every morning numerous females, from all parts of the neighborhood, bring for sale into Seringapatam baskets of this fuel. Many females who carry large baskets of cow-dung on their heads are well-dressed and elegantly formed girls.

In another place Bourke speaks of "an English writer" of 1832 who describes "remnants of the Hindu sect of the Aghozis" as follows:

> In proof of their indifference to worldly objects they eat and drink whatever is given to them, even ordure and carrion. They smear their bodies also with excrement, and carry it about with them in a wooden cup, or skull, either to swallow it, if by so doing they can get a few pice [coins], or to throw it upon the persons or into the houses of those who refuse to comply with their demands. . . . The Abbé Dubois says that the Gurus, or Indian priests, sometimes, as a mark of favor, present to their disciples "the water in which they had washed their feet, which is preserved and sometimes drunk by those who receive it. . . . Neither is it the most disgusting of the practices that prevail in that sect of fanatics, as they are under the reproach of eating as a hallowed morsel the very ordure that proceeds from their Gurus, and swallowing the water with which they have rinsed their mouths or washed their faces, with many other practices equally revolting to nature. . . ."

Perhaps the most remarkable example of dung-worship concerns the Grand Lama of Tibet, about whom Bourke has collected information from several sources. The Jesuit Gruebner, who walked through Tibet in 1661, is thus cited from Pinkerton (1814):

> The grandees of the kingdom are very anxious to procure the excrements of . . . the Grand Lama . . . which they usually wear about their necks as relics. . . . The Lamas make a great advantage

of the large presents they receive for helping the grandees to some of his excrements, or urine; for by wearing the first about their necks, and mixing the latter with their victuals, they imagine themselves to be secure against all bodily infirmities. In confirmation of this, Gerbillon informs us that the Mongols wear his excrements, pulverized, in little bags about their necks, as precious relics, capable of preserving them from all misfortunes, and curing them of all sorts of distempers. . .

And again:

Warren Hastings speaks of the Thibetan priests of high degree, the "Ku-tchuch-tus," who, he says, "admit a superiority to the Dalai Lama, so that his excrements are sold as charms, at great price, among all the Tartar tribes of this religion."

Bourke suggests that the Grand Lama was not above using a little chicanery in this transaction. Holy materials called "Pedung pills" as well as by other names appear to have been made from grain rather than from the Lama's feces, but with an elaborate ceremony that converted them into a "symbolical alvine dejection." The suggestion is made that

the most plausible explanation is, that the lamas, finding trade good and the Buddhist laity willing to accept more "amulets" than the Grand Lama was able, unaided, to supply, hit upon this truly miraculous mode of replenishing their stock.

Bourke relates the oriental word *pedung* to words similar to *dung* in all the Aryan languages.

Bourke got hold of four of the sacred pills and sent them to the United States Army laboratories for analysis. He presents a letter in response from W. H. Mew, dated April 18, 1889, which reports demurely that "the Grand Lama's ordure" contained a good deal of undigested starch, as well as what looked like wheat cellulose, suggesting that "the flour used . . . was of a coarse quality, and probably was not made in Minnesota." The composition was otherwise unremarkable, showing a slight reaction for biliary matter, suggesting no obstruction of the Lama's bile ducts.

Vestiges of dung-worship appear to have intruded into early Christian practices. Bourke devotes a chapter, which I found wordy and unclear, to the "Stercoranistes" or "Stercorians," who he says were not dealt with in editions of the *Encyclopædia Britannica* after that of 1841, from which he quotes:

formed from *stercus*, "dung," a name which those of the Romish church originally gave to such as held the Host was liable to digestion and all its consequences, like other food.

These practices were evidently perverted relatives of those of the Eucharist and Corpus Christi, mixed with pagan vestiges, among the illiterates of feudal times, during the fifth and sixth centuries, and in which miracles

> were wrought either by the swaddling clothes themselves or by the water in which they had been cleansed; and the inference is that the excreta of Christ were believed, as in many other instances, to have the character of a panacea, as well as generally miraculous properties.

The following appears in Richard Burton's *Arabian Nights* (Volume 2). King Afridun is addressing the Emirs:

> ". . . And I purpose this night to sacre you all with the Holy Incense." When the Emirs heard these words they kissed the ground before him. Now the incense which he designated was the excrement of the Chief Patriarch, the denier, the defiler of the Truth, and they sought for it with such instance, and they so highly valued it that the high priest of the Greeks used to send it to all the countries of the Christians in silken wraps after mixing it with musk and ambergris. Hearing of it Kings would pay a thousand gold pieces for every dram and they sent for and sought it to fumigate brides [*sic*] withal; and the Chief Priests and the great Kings were wont to use a little of it as collyrium for the eyes and as a remedy in sickness and colic; and the Patriarchs used to mix their own skite with it, for that the skite of the Chief Patriarch could not suffice for ten countries . . .

Whereupon, without further detail, Shahrazad, perceiving the dawn, and being thus saved for another day, breaks off, to resume the following [ninetieth] night, after the customary preliminaries, with:

> King Afridun summoned his chief Knights and Nobles and . . . incensed them with the incense which as aforesaid was the skite of the Chief Patriarch . . . This incensing done, he called for Luka bin Shamlut, surnamed the Sword of the Messiah; and, after fumigating him and rubbing his palate with the Holy Merde, caused him to snuff it and smeared his cheeks and anointed his moustaches with the rest . . .

So it would appear that cleverness is not the monopoly of any one time or place. Musk and ambergris are, of course, both animal secretions. As for "Holy Merde," we know that ideas of curative value have not been limited to any such variety of the general product.

Let me insert here an item suggesting that the early Roman church (late eleventh to early twelfth centuries) could be explicit in malediction (in Latin) if not otherwise. Such seems to be the import of the curse of excommunication as formulated by Bishop Ernulphus of Rochester, as Laurence Sterne gives it in full both in Latin and in English in *Tristram Shandy*. It was being read aloud by Dr. Slop as Uncle Toby whistled "Lillabullero." I give the relevant paragraph only, in both languages:

> *Maledictus sit viviendo, moriendo* [Sterne inserts a series of dashes here] *manducando, bibendo, esuriendo, sitiendo, jejunando, dormitando, vigilando, ambulando, stando, sedendo, jacendo, operando, quiescendo, mingendo, cacando, flebotomanso.*

> May he be cursed in living, in dying . . . in eating and drinking, in being hungry, in being thirsty, in fasting, in sleeping, in slumbering, in walking, in standing, in sitting, in lying, in working, in resting, in pissing, in shitting, and in blood-letting.

A number of practices given by Bourke are distinctly reminiscent of sympathetic magic. Feces, urine, and other products of man and animals were used as aphrodisiacs and antiphrodisiacs or for less explicit purposes usually related to courtship and marriage.

Uncomplicated sympathetic magic is illustrated in this quotation, attributed by Bourke to the Reverend Chatelain, on customs in Angola, West Africa.

> When a young man is trying to win the love of a reluctant girl he consults the medicine-man, who then tries to find some of the urine and saliva which the girl has voided, as well as the sand upon which it [*sic*] has fallen. He mixes them with a few twigs of certain woods, and places them in a gourd, and gives them to the young man, who takes them home, and adds a portion of tobacco. In about an hour he takes it out and gives it to the girl to smoke; this effects a complete transformation in her feelings.

Bourke mentions that the same ingredients are used in an antiphrodisiac or for related purposes. Some of these, like other prescriptions we shall come to mention, were in vogue in Europe in relatively modern times, as well as among peoples we like to think of as barbarians. The following example is attributed to Schurig (1725):

> . . . a man who wore for an hour shoes in which had been put the excrement of his lady love "was completely cured of his infatuation."

The following from *Wit without Money* by Beaumont and Fletcher seems to have similar significance:

> RALPH: Pray, empty my right shoe, that you made your chamber-pot, and burn some rosemary in it.

A related item from Bourke is given as follows:

> In the *Histoire Sécrète du Prince Croq'Êtron,* "M'lle. Laubert, Paris, 1790, Prince Constipati is entertained by the Princess Clysterine; elle lui donne de la limonade, de la façon d'urinette."

Semen and menstrual fluid were used as aphrodisiacs. Bourke says this of semen:

> There is nothing to show whether male lovers used this ingredient and maidens the menstrual liquid, or both indiscriminately; but it seems plausible to believe that each sex adhered to its own excretion.

He nevertheless attributes the following to Samuel Augustus Flemming, in a treatise on remedies from human sources dated 1738:

> From semen was prepared "what was known as 'magnetic mummy,' which, being given to a woman threw her into an inextinguishable frensy of love for the man or animal yielding it."

It is fair to record that Captain Bourke expresses horror at this and other suggestions in Flemming of the participation of animals in such matters. At the same time he notes, among other aphrodisiacs, leopard dung and the urine of a bull voided immediately after the bull has covered (copulated); the urine was "taken in drink," or "the groin [was] well rubbed with earth moistened with this urine." Also, an

> ointment of the gall of goats, incense, goat-dung, and nettle-seeds was applied to the privy parts previous to copulating to increase the amorousness of women.

Paracelsus is quoted as having taught that

> when one person ate or drank anything given off by the skin of another, he would fall desperately in love with that other.

And from one Daniel Beckherius (1660) comes an allusion to the use in love philters of sweat as well as of menstrual fluid or semen.

Another item, combining sweat and shoes, taken from Flemming, is this:

> It was narrated that if a man, who under the influence of a philtre, was forced to love a girl against his will, would put on a pair of new shoes, and wear them out by walking in them, and then drink wine out of the right shoe, where it could mingle with the perspiration already there, he would promptly be cured of his love, and hate take its place.

Sympathetic magic is again suggested in several items given by Bourke in much the form of this one:

> Pisse through a wedding-ring if you would know who is hurt in his privities by witchcraft.

This procedure was suggested as a cure for impotence among other purposes. Bourke says this remedy was believed in and known to have been practiced by the peasantry in some parts of Germany at least as recently as 1861.

11

Husbandry on Earth

The salt of man's urine hath an excellent quality to cleanse . . .
Man's dung, or excrement, hath very great virtues, because it
contains in it all the noble essences, viz.: of the Food and Drink.

—Paracelsus

Before abnegation changed man's early attitude toward
his excretions into the one we have today, he felt awe and reverence
toward them. He cherished them for real value tangibly demonstrated.
The stuff was useful, and therefore the more magical. Feces renewed
the life-giving virtues of the earth; urine did a better job of cleaning
than plain water. Before man got rich and prodigal, such gifts were
not to be denied; and since they were wonderful and mysterious it
seemed plain that the same things must work other magic as well.
Today it is hard to reach back through our cultivated aversion and our
affluence, when these materials have come to be prime symbols of
worthlessness, to a time when aversion had not been thought of and
anything useful was necessarily treasured. But reach back is just what
we must do now. If we can keep a steady stomach, we may see that
some of what the ancients learned and we have forgotten may be
worth relearning. Their empiricism brought forth a tangle of truth
and fantasy. Science can now explain the truth and teach us to throw
away the rest. But we must pry away the crust of prejudice.

We do not know exactly how or where the practice of manuring
began, or whether human excrement was used for it before animals
were herded in sufficient numbers to meet the need. Agriculture and
animal husbandry started together in Neolithic times, no doubt in
many separate places. The earliest fully accepted evidence of manuring comes from the Neolithic Swiss lake dwellings, where the practice

was evidently already old and well established. Here the provider was the small cow of the Lascaux paintings. Primitive scientists must have tested and verified the observation that naturally manured ground was especially fertile; and in the absence of prejudice it seems likely that, if indeed the first observations in some areas were not made with human excrement, they would soon have been transferred to that product by a scientific method involving what may have been the first animal experiments. We know that the use of human manure in agriculture by the Chinese continued down to recent times, and that compost, in which human and animal dung and urine were mixed with decaying vegetable material, was in very early use there. These practices were evidently associated with the most primitive religious ideas, expressed in fertility rites and dung gods.

Augeus is said to have been the first king among the Greeks to use human dung for the fields. He reported that "Hercules divulges the practice thereof among the Italians." But in the story it was of course not human but cattle dung in the Augean stables, the dung of thousands of cattle, that Hercules cleared away as his fifth labor.

Urine must have been husbanded before historical times, but the earliest reference I have come across relates to Vespasian in the days when he was aedile, or official in charge of public works, before becoming emperor of Rome. Harington tells how he increased the city's revenues:

> But afterward himselfe coming to be Emperour (though the citie of Rome was before his time sufficiently furnished of Jaxes [jakes, privies]) yet it seemed there wanted other places of neare affinitie to them (which he found belike when he was Aedile by experience) I meane certain pissing conduites: and therefore he caused diverse to be erected in the most populous and frequented places of the Citie, and saved all the urine in cisternes, and sold it for a good summe of money to the Dyers . . . When his sonne Titus seemed to finde fault with him for devising a kind of tribute, even out of urine; the monie that came into his hand of the first paiment, he put into his sones nose: asking withall, whether he was offended with the smell, or no, and when answered no: "And yet," quoth he, "it cometh of Urine."

To which the aphorism is appended:

> So we get the chinks,
> We will bear with the stinks,

together with the suggestion that Vespasian be deified as "Urinatus, of *Urina*, like Stercutius, of Stercus." Harington's editor provides a footnote here: "A water fountain in London located near the Royal Exchange, characterized by its small stream of water, was called the pissing conduit." In Part II of *Henry VI*, Shakespeare has the briefly victorious Jack Cade say:

> . . . And here sitting upon London Stone, I charge and command, that of the City's cost the pissing-conduit run nothing but claret wine this first year of our reign . . .

Another version of the Vespasian story is given by Bourke from Lucretius, as follows, translated from the Latin by J. M. Good in 1805:

> Urinary reservoirs were erected in the streets of Rome, either for the purpose of public cleanliness, or for the use of the fullers, who were accustomed to purchase their contents of the Roman government during the reign of Vespasian, and perhaps other emperors, at a certain annual impost, and which, prior to the invention or general use of soap, was the substance employed principally in their mills for cleansing cloths and stuffs previous to their being dyed.

The following appears in a work by Eugene O'Curry on *The Manners and Customs of the Ancient Irish* (1873):

> The preparation of blue, violet and bluish-red coloring matters from lichens by the action of the ammonia of stale urine, seems to have been known at a very early period to the Mediterranean peoples, and the existence, down almost to the present day, of such a knowledge in the more remote parts of Ireland, Scotland, and Scandinavia, renders it not improbable that the art of making such dyes was not unknown to the northern nations of Europe also.

Havelock Ellis is quoted from a personal letter to Bourke as saying of urine that

> nearly everywhere it has been the first soap known. . . . In England and France, and probably elsewhere, the custom of washing the hands in urine, with an idea of its softening and beautifying influence, still subsists among ladies, and I have known those who constantly made water on their hands with this idea.

The Hindu, venerating the cow, used her urine for washing and "for removing all imaginable uncleanness" (Dubois, 1817). Bourke mentions in various places that urine was used for washing and related purposes by numerous American Indians, Mexicans, and Africans, as well as by natives of Spain. The urine was allowed to stand until ammonia formed in it and was then used not only for ablutions in general but as a mouthwash. This custom was transplanted by the Spanish colonists to America and persisted in Florida in the mid-nineteenth century. Human or animal urine was also valued as a source of salt by such diverse peoples as Egyptians and other North Africans, Hindus, the Indians of Bogatá, the Buryates of Siberia, and the German settlers in Pennsylvania. According to the *Encyclopedia* of Diderot and d'Alembert (1789), the sal ammoniac of the ancients was prepared from the urine of camels; while phosphorus and saltpeter, as then manufactured in England, were made from human urine. What appears to have been a flux for soldering gold, called by Galen "chrysocollon" (gold-glue), was prepared, Bourke tells us,

> from the urine of a boy, who had to void it into a mortar of red copper while a pestle of the same material was in motion, which urine was carefully exposed to the sun until it had acquired the thickness of honey.

Urine has also been used to remove ink spots (Pliny), and among Eskimos and Indians as the vehicle in tattooing by which carbon dust was introduced into the skin.

It appears that urine was used for washing in preference to water even when water was freely available; but among Eskimos and other northern peoples urine evidently enjoyed a special place in custom. The following story about the Siberian Chukchee ("Tchoukchi") is of special interest both for its evidently authentic detail and for its unusual sympathy for the people described. It is quoted by Bourke as a personal letter to him from William H. Gilder, author of *Ice-Pack and Tundra* (1883), among other works. The letter was dated New York, October 15, 1889.

> . . . the food is served in the "yoronger," or inner tent, in which men and women sit, in a state of nudity, wearing only a small long-cloth [*sic*] of seal-skin. After finishing the meal, "a small, shallow pail or pan is passed to any one who feels so inclined, to furnish the warm urine with which the board and knife are washed by the housewife. It is a matter of indifference who furnishes the fluid, whether the men, women, or children; and I have

myself frequently supplied the landlady with the dishwater. In nearly every tent there is kept from the summer season a small supply of dried grass. A little bunch of this is dipped in the warm urine and serves as a dish-rag and a napkin. These people are generally kind and hospitable, and were very attentive to my wants as a stranger, and regarded by them as more helpless than a native. The women would, therefore, often turn to me after washing the board and knife, and wash my fingers and wipe the grease from my mouth with the moistened grass. Any of the men or women in the tent who desired it would also ask for the wet grass, and use it in the same way. It was not done as a ceremony, but merely as a matter of course or of necessity. I do not think they would use urine for such purposes if they could get all the water, and especially the warm water, they needed. But all the water they have in winter is obtained by melting snow or ice over an oil-lamp,—a very slow process: and the supply is therefore limited [to] drinking purposes, or to boil such fresh meat as they may have. The urine, being warm and containing a small quantity of ammonia, is particularly well adapted for removing grease from the board and utensils, which would otherwise soon become foul, and to their taste much more disagreeable. The bottom of the 'yoronger' is generally carpeted with tanned seal-skins, and they too are frequently washed with the same fluid. The consequence is that there is ever a mingled odor of ammonia and rotten walrus-meat pervading a well-supplied and thrifty Tchouktchi dwelling."

Here is another of Bourke's sources (Cochrane, 1883), evidently speaking of the same people under the same circumstances:

Their stench and filth are inconceivable. . . . The large tents were disgustingly dirty and offensive, exhibiting every species of grossness and indelicacy. . . . It would be impossible, with decency, to describe their habits, or explain how their very efforts toward cleanliness make them all the more disgusting.

Such is the eye of the beholder.

Urine has a place in the ceremonial steam-baths of Alaskan Eskimo men. The ceremony is held in the *kashim*, or *kashga*, the large circular smokehouse built partly below ground so that only a low mound shows from above. It is approached through a tunnel, and is open otherwise only through a smokehole in the center of its roof. The Eskimos group themselves inside around a central fire built on stones. Here is a description of the ceremony as given in Bourke by Henry W. Elliot (1887):

> During the kashim ceremony, a native "performed the disgusting operation of urinating over the back and shoulders of the person seated next to him, after which he jumped down and began to dance. . . . The one urinated on then repeated the operation on his next neighbor, and this continued until the last man urinated on the first. . . . The fire is usually drawn from the hot stones on the hearth. . . . A kantog of chamber-lye poured over [the hot stones] which, rising in dense clouds of vapor, gives notice by its presence and its horrible ammoniacal odor to the delighted inmates that the bath is on. The kashga is heated to suffocation; it is full of smoke; and the outside men run in from their huts with wisps of dry grass for towels and bunches of alder twigs to flog their naked bodies.
>
> "They throw off their garments; they shout and dance and whip themselves into profuse perspiration as they caper in the hot vapor. More of their disgusting substitute for soap is rubbed on, and produces a lather, which they rub off with cold water. . . . This is the most enjoyable occasion in an Indian's existence, as he solemnly affirms."

This procedure, except for the substitution of urine for water, including the final cold-water rinse, is reminiscent of Russian and Turkish steam bathing. The cold water was presumably brought into the *kashga* as buckets or containers of melting snow or ice. Elliot's mention of "Indians" may be an error; but Bourke records that *kashga*-like houses were used by the Pueblos, as well as in Siberia, and that the model has been perpetuated in the temples of India. In 1885, George Kennan, in his *Tent Life in Siberia*, suggested that the roof-opening in the Siberian smokehouse may be associated with the origin of the Santa Claus story; it would certainly have been more accessible than the chimney.

Human and animal feces had other uses than as manure. In addition to the use of cow dung as fuel by the Hindus, mentioned in the previous chapter, Bourke speaks of the use of this and other animal dung as plaster or mortar for walls and as paving for floors, especially in the ancient East and in Africa, as well as in Egypt as recently as 1869. He also adds substance to the rumor I had heard a long time ago and dismissed as merely scurrilous: that excrement has been used in the preparation of tobacco. He quotes the following from Schurig (1725):

> The best varieties of tobacco coming from America were arranged in bunches, tied to stakes, and suspended in privies, in order that the fumes arising from the human ordure and urine might correct the corrupt and noxious principles in the plant in the crude state.

A letter from Dr. Gustav Jaegar, dated August 29, 1888, is quoted as saying, "I heard lately from good authority that, in Havana, the female urine is used in cigar manufacturing as a good maceration." And elsewhere Bourke adds:

> Father DeSmet says of the Flathead and Crow Indians (1846): "To render the odor of the pacific incense agreeable to their gods, it is necessary that the tobacco and the herb (skwiltz), the usual ingredients, should be mixed with a small quantity of buffalo dung."

Similar customs appear to have prevailed among the Sioux, Cheyennes, Arapahoes, and other Plains Indians, "to whom the buffalo is a god." Hottentots, "when in want of tobacco," we learn, "smoke the dung of the two-horned rhinoceros or of elephants." In Kipling's story, "Miss Youghal's Sais," Strickland, the Englishman in India who becomes Miss Youghal's *sais*, or groom, for love of her, "learned to smoke tobacco that was three-fourths dung" among other native customs in the course of winning her as his wife.

Bourke, speaking of the adulteration of opium with hen manure, was prompted to draw the curious conclusion that the smoker of opium

> is thus placed on a par with the American Indian smoking the dried dung of the buffalo, and the African smoking that of the antelope or the rhinoceros.

We also learn in this context that the odor of musk and the color of coral were restored by suspending them in a privy for a time.

Another commentator on the Chukchee of Siberia (Dulaire, 1825), whose friendly customs we have already looked into, is cited for the following item, given by Bourke for unstated reasons in French. I have taxed my poor knowledge of that language to translate. These people

> offer their women to travelers; but these latter, to be shown worthy, must submit to a disgusting test. The girl or woman who

is to pass the night with her new host gives him a cup full of her urine; he must rinse his mouth with it. If he has the courage to do so he is looked upon as a sincere friend; if not, he is treated as an enemy of the family.

A remarkable group of stories, the most notable of which I have taken pains to authenticate, deals with intoxicating food or drink and includes the use for this purpose of urine. Bourke says an edible pine nut is found in Queensland, Australia,

> of which the natives are extremely fond. . . . The men would form large clay pans in the soil, into which they would urinate; they would then collect an abundance of these seeds and steep them in the urine. A fermentation took place, and all the seeds were devoured greedily, the effect being to cause a temporary madness among the men—in fact a perfect delirium tremens.

In this instance the liquid was discarded, but in others the urine itself was imbibed. The medicine men of the Cape Flattery Indians in what is now British Columbia

> distill from potatoes and other ingredients, a vile liquor, which has an irritating and exciting effect upon the kidneys and bladder. Each one who partakes of this dish immediately urinates and passes the result to his next neighbor, who drinks. The effect is as above, and likewise a temporary insanity or delirium, during which all sorts of mad capers are carried out. The last man who quaffs the poison, distilled through the persons of five or six comrades, is so completely overcome that he falls in a dead stupor.

The speed of this "distillation" is improbable, but independent testimony points to a similar transmission chain via urine. This brings us to the tangled tale of the poisonous mushroom *Amanita muscaria*.

This mushroom has been known for centuries as the fly fungus or fly agaric, having been used to poison house flies. Another use was first documented, as far as I have traced it back, by Oliver Goldsmith— that is, as a promoter of conviviality, like alcohol as we use it today, but with the special added property of being excreted unchanged in the urine. I found it strange that the current medical literature in the relevant fields of pharmacology and toxicology, as far as I have looked into it, says nothing whatever on this subject. The mystery is all the greater in that *Amanita muscaria* is treated prominently and at length as a highly poisonous mushroom from which the potent alkaloid muscarine is derived. Could this be another instance of delicate expurgation?

The story opens in 1762 with the thirty-second letter of Goldsmith's *Citizen of the World*, subtitled *Letters from a Chinese Philosopher Residing in London to his Friends in the East*. In its course a footnote gives a brief reference to one O. G. Van Stralenberg as having provided similar information: I have not explored this lead. Goldsmith's story is narrated by his English character, the "Man in Black," who has spoken of the difficulty of an "English nobleman to preserve his appearance of greatness," whereupon the Chinese responds with this description of something he has seen in his travels:

> "The pitiful humiliation of the gentleman you are now describing," said I, "puts me in mind of a custom among the Tartars of Korecki not entirely dissimilar to this we are now considering. The Russians, who trade with them, carry thither a kind of mushrooms, which they exchange for furs of squirrels, ermines, sables, and foxes. These mushrooms the rich Tartars lay up in large quantities for the winter, and when a nobleman makes a mushroom feast all the neighbors around are invited. The mushrooms are prepared by boiling, by which the water acquires an intoxicating quality, and is a sort of drink which the Tartars prize beyond all other. When the nobility and the ladies are assembled, and the ceremonies usual between people of distinction over, the mushroom-broth goes freely round; they laugh, talk, *double entendre*, grow befuddled, and become excellent company. The poorer sort, who love mushroom-broth to distraction as well as the rich, but cannot afford it at the first hand, post themselves on these occasions round the huts of the rich, and watch the opportunity of the ladies and gentlemen as they come down to pass their liquor; and, holding a wooden bowl, catch the delicious fluid, very little altered by filtration, being still strongly tinctured with the intoxicating quality. Of this they drink with the utmost satisfaction, and thus they get as drunk and as jovial as their betters."

The Man in Black adds a satirical rejoinder, placing the English nobility in a descending series, the wooden bowl serving as analogy for an equivalent obsequious gesture at each level to the one above.

The following item is cited by Bourke from the *English Cyclopædia* of 1854, article "Fungi":

... [a] variety of *Amanita muscaria* is used by the inhabitants of ... northeastern ... Asia ... as wine, brandy, arrack, opium, etc., is by other nations ... collected in the hottest months ... hung up ... to dry. ... It is said that from time immemorial the inhabitants have known that the fungus imparts an intoxicating quality to [urine] ... which continues for a considerable time after taking it. For instance, a man moderately intoxicated today will by the next morning have slept himself sober; but (as is the custom) by taking a cup of his urine he will be more powerfully intoxicated than he was the previous day. It is therefore not uncommon for confirmed drunkards to preserve their urine as a precious liquor against a scarcity of the fungus ... thus with a very few *Amanitae* a party of drunkards may keep up their debauch for a week.

As I said before, information on *Amanita muscaria* is plentiful in the places where one would naturally go for verification of these fantastic tales—the texts of pharmacology and toxicology. But the matter of urine-drinking is so conspicuously absent in these books that I would have given the story differently, possibly have omitted it, had I not found what I was looking for in other places. In the medical texts it appears that *Amanita muscaria*, besides being the source of muscarine, is neither so common nor so poisonous as the related mushroom *Amanita phalloides*. But this latter has nothing to do with the case except, perhaps, to suggest that the imagery of taxonomic botanists is relatively unfettered, and to confirm our suspicion that it is all right to say almost anything in Greek. Muscarine comes from the fly fungus itself, and leads into matters fundamental to neurophysiology, such that a group of substances are called "muscarinic." Two others besides muscarine itself are included—arecoline from the betel nut, and the single useful drug among the three, pilocarpine, which promotes the flow of saliva and tears. One turns up other curious intelligence: that the wife and three children of Euripides (described innocently as a Greek poet, fifth century B.C.) died from eating poisonous mushrooms, as did the Czar Alexis in 1676. There is a paper by W. W. Ford, dated 1923, in which five varieties of "mycetismus," or mushroom poisoning, are listed, of which the last and evidently the least is "mycetismus cerebralis, with transient excitement and hallucination." At this point, having found not a word on urine-drinking, I abandoned this particular approach and looked for another one.

If medicine has forgotten Oliver Goldsmith—although he was a

physician himself!—and if urine-drinking is a subject that Nice
People do not talk about, or whatever the reason may be, the literature
of psychedelic agents has picked up the trail, and from it I was led
to what I was looking for. But at first the scent was no more than
tantalizing. I ran across one item in the so-called underground press,
in the San Francisco *Oracle*, Volume 1, Number 7, 1967. It is mainly
concerned with the "magic mushrooms" of the Aztecs, which were
described by a banker named R. Gordon Wasson in an article in *Life*
magazine in 1957. Here we get off the track and among mushrooms
unrelated to *Amanita*, mushrooms with analeptic or ataraxic properties
—that is to say, producers of hallucinations, psychedelic agents proper.
The author of the *Oracle* story, Gene Grimm, passes rapidly from
these through the banana-peel agent to the *Amanitae*, of which he
speaks with a profundity I found unconvincing. He says this, suggesting a Hindu source but not giving it:

> The toxic effects of A. muscaria can be partially eliminated by
> removing the skin and warts of the caps, or by marinating the
> mushroom in salt or vinegar, or by drinking milk when the mush-
> room is ingested (these are traditional means of making the
> mushroom safe for ingestion).

Among other current sources in which some of this material is repeated with varying levels of credibility, I was impressed with a little
book by Norman Taylor, a botanist known for his works on gardening.
He devotes a section to the fly agaric and gives some firm information.
This mushroom, we learn, grows in the forests and forest-edges of
New York State, where it has a whitish stalk swollen at the base, a
lacerated collar about three-quarters of the way up the stalk, and a
beautifully colored cap like an umbrella (the "pileus") that may be as
much as eight inches in diameter. The pileus of North American
varieties is whitish, yellowish, or orange-red, but in Europe and Asia
it is bright red or purple. He cites the H. G. Wells story called "The
Purple Pileus," in which the hero, reacting to a shrewish wife, attempts
suicide by eating fly agaric, but eats only enough to get roaring drunk.
Taylor goes on to speak—with aversion if not with horror—of urine-
drinking among the Chukchee. He suggests that their special circumstances compel them to do so, and urges compassion.

> Perhaps it helps to sustain them on long reindeer hunts over a
> bleak, cheerless terrain—perhaps the most bitter in the world.
> Who are we to deny them this revolting pleasure? To these dull

plodders of Arctic wastes the fly agaric may well be their only peep into a world far removed from the frozen reality of their wretchedness.

These words bear comparison with accounts of the same people by Gilder and by Cochrane, which we saw before. The observer evidently tends to include himself in his picture.

An eyewitness of "toadstool drunkenness"—as he called it—was George Kennan, whose *Tent Life in Siberia* I mentioned before. Bourke cites the edition of 1885. In a later edition (1910) I found a direct account of the practice among the "Koraks" (Koryaks) of Kamchatka. The fungus, known locally as "muk-a-moor," is identified as "*Agaricus muscarius*," or fly agaric. Its use was prohibited by Russian law; and since it does not grow on the local "barren steppes" the Koryaks bought it from trading bootleggers at enormous prices: a single fungus might bring twenty dollars' worth of furs. But a whole band could get drunk on a single mushroom. Kennan could not force himself to be explicit. He says:

> For a description of the manner in which this band gets drunk collectively and individually upon one fungus, the curious reader is referred to Goldsmith's *Citizen of the World*, Letter 32. It is but just to say, however, that this horrible practice is almost entirely confined to the settled Koraks of Penzhinsk Gulf—the lowest, most degraded portion of the whole tribe. It may prevail to a limited extent among the wandering natives, but I have never heard of more than one such instance outside of the Penzhinsk Gulf settlements.

The most convincing account of these goings-on is that of Waldemar Bogoras, a naturalist whose treatise *The Chukchee* (1904) contains the following excerpt:

> Fly agaric is the only means of intoxication discovered by the natives of northeastern Asia. Its use is more common in the Koryak tribe, as agaric does not grow outside of the forest border. For the same reason only the [southernmost] Chukchee—e.g., those around the Anadyr, Bug River, and Opuka River—are supplied with the intoxicating mushroom. They do not compare with the Koryak, however, in their passion for agaric . . .

Then, after details of the way the mushrooms are prepared and eaten, and of the symptoms that follow, which include hallucinations as well as excitement and joviality, we find this paragraph, which is all Dr. Bogoras has to say on the subject:

> Drinking the urine of one who has recently eaten fly agaric produces the same effect as eating the mushroom. The passion for intoxication becomes so strong that the people will often resort to this source when agaric is not available. Apparently without aversion they will even pass this liquor around in their ordinary teacups. The effect is said to be less than from the mushrooms themselves.

All this testimony, it seems to me, forces us to conclude that the Chukchee people of Siberia use human urine as a detergent for their utensils and their lips, offer it as a mouthwash in a test of friendship, and drink it for its effects when it contains the intoxicating principle of the fly agaric. The whole record indicates that in doing these things they show no preference for the urine as such over the alternatives of warm water or the mushroom itself; in this respect they seem more fastidious than Havelock Ellis's ladies and many another people. But they also have no aversion to these uses of urine, although their friendship test shows an awareness of aversion in others. The account is also full of manifestations of revulsion on the part of these others, witnesses to the events; and we would ordinarily identify ourselves more with these others than with the Chukchee. But our revulsion, our occasional commiseration, our contempt or pity, surely reflect more on ourselves than they do on the Chukchee!

Other examples of the use of urine and feces, directly or indirectly, as food or drink are scattered through the pages of Bourke. Some are associated with insanity, and these I omit. Others deal with dire necessity, or with duress. Examples of necessity include Biblical references (2 Kings 18:27; Isaiah 36:12; Ezekiel 4:12-15) to dung-eating and urine-drinking among the Israelites, and the use of urine in the prolonged absence of drinking water in "the sieges of Jerusalem, Numantia, Ghent, the famine in France under Louis XIV," and many other instances, including one of shipwrecked English soldiers, and another in 1877 when the command of Captain Tolan, scouting after Indians in Texas,

> was reduced to living for several days on the blood of their own horses and their own urine, water not being discovered in that vicinity.

(In Peter Weiss's play *The Investigation*, an inmate of Auschwitz, condemned to death by starvation in solitary confinement, drank his own urine.)

The chickpea, or garbanzo, is thought to be the "dove's dung" mentioned in relation to the siege of Samaria in 2 Kings 6:25: "the fourth part of a cab of dove's dung [sold] for five pieces of silver." This so-called chichi (*pois chiches*), being pointed at one end and assuming an ashen color when parched, is said to be still called dove's dung by the Arabs.

The presence of dung in sausage or other food prepared from animal entrails is suggested in a letter sent by the wife of the Elector of Hanover in 1694. Bourke tells us this

> may serve to give an idea of the boldness of the opinions enter-
> tained by the ladies of high rank in that era, and the coarseness
> with which they expressed themselves.

I give the passage this time in French as Bourke gives it, followed by my attempt at translation, pleading that freedom of speech as it is guaranteed to us specifies no condition of a foreign language:

> *Si la viande fait la merde, il est vrai de dire que la merde fait
> la viande. . . . Est-ce que dans les tables les plus délicates, la
> merde n'y est pas servie en ragoûts? . . . Les boudins, les an-
> douilles, les saucisses, ne sont-ce pas de ragoûts dans des sacs
> à merde?*
>
> If meat is shit, it is correct to say that shit is meat. . . . Isn't
> it true that on the daintiest tables shit is served in stews? . . .
> Black-puddings, chitterlings, sausage, are they not stews in shit-
> bags?

The following is given by Bourke as a quotation attributed by Herbert Spencer to Lewis and Clark:

> One of them [evidently an American Indian], who had seized
> about nine feet of the entrails, was chewing it at one end, while
> with his hand he was diligently clearing his way by discharging
> the contents of the other.

The "black pudding" of Samuel Butler's *Hudibras*, the *boudins* of the Feast of Fools, and other sausage, were made from the flesh, blood, and entrails of pork killed by several families in common on December 17, known as "Sow Day."

The "flapdragon" of Elizabethan times seems to refer to drinking one's health in urine. It appears in Marston's *Dutch Courtesan,* in Dekker's *Honest Whore,* Part I, and in Middleton's *A Trick to Catch the Old One.* Bourke speaks of a footnote in an 1825 edition of Dekker that includes the words "Dutch flap-dragons . . . healths in urine." An edition I have seen that includes both Dekker plays and that of Middleton explains "flapdragon" in a footnote as "a raisin floating on burning brandy"; but in the Middleton play the hero, Witgood, in his confession at the very end, disclaims

> Dutch flapdragons, healths in urine,

to which a footnote adds

> Dutchmen had the reputation of being very expert in swallowing flapdragons (Bullen).

A paper by Richard Neal, M.D., in the medical journal *Practitioner* (London) for November 1881 (page 343) makes this statement in connection with the medicinal use of urine in South America:

> As a stimulant and general pick-up, I have frequently seen a glass of child's or a young girl's urine tossed off with great gusto and apparent benefit.

In contrast to the use of excrement as food in dire necessity, as by the Israelites, are three gourmet items. The first comes from an article entitled "The Last Cholera Epidemic in Paris" in the *General Homeo-pathic Journal,* Volume 113, page 15 (1886):

> The neighbors of an establishment famous for its excellent bread, pastry, and similar products of luxury, complained again and again of the disgusting smells which prevailed therein and which penetrated into their dwellings. The appearance of cholera finally lent force to these complaints, and the sanitary inspectors who were sent to investigate the matter found that there was a con-nection between the water-closets of these dwellings and the reser-voir containing the water used in the preparation of the bread. This connection was cut off at once, but the immediate result thereof was a perceptible deterioration of the quality of the bread. Chemists have evidently no difficulty in demonstrating that water impregnated with "extract of water closet" has the peculiar prop-erty of causing dough to rise particularly fine, thereby imparting to bread the nice appearance and pleasant flavor which is the principal quality of luxurious bread.

The second item is from a letter written to Bourke in 1888 by Dr. Gustav Jaeger:

> A storekeeper in Berlin was punished some years ago for having used the urine of young girls with a view to make his cheese richer and more piquant. Notwithstanding, people went, bought and ate his cheese with delight.

And another letter from a Dr. Bernard, dated the same year, mentions having heard that certain Swiss firms used urine to activate cheese fermentations.

A bacteriologist would anticipate that by now appropriate strains of microbic species contributing these effects would have been isolated for use in pure culture apart from cholera vibrios and other unpleasantness, as well as from the remainder of the natural vehicle, which would not be expected to add anything significant to the desired reaction. But it remains true that advances in fermentation technology, great as they have been in recent years, have not as yet fully supplanted older and not necessarily scientific methods of producing wine and cheese. The problem, however, certainly entails nothing more subtle than the exactly right microbe or microbes and the proper conditions for their growth, and I would expect it to be solved in time.

At this point, if you were not aware of it before, you will not be surprised to learn that feces and urine, as well as other excretory and secretory products of man and animals, were once extensively used by physicians in the treatment of disease. An up-to-date treatment of this subject by an informed non-physician, giving the doctor his due both as rational gentleman and as pompous humbug, would be a good idea; but I have no compulsion to undertake it. It has always been true, I think, and still is even today, when physicians think of themselves as scientists—perhaps precisely because they do—that they have needed and used the arts of magic in their trade. The nineteenth century saw an awakening of scientific therapeutics, but the twentieth, largely as a result of successes based on the preceding victories of science, has witnessed a proliferation of new and baffling problems. It seems to me that the prescription of drugs is again, sometimes, based on a symptomatic empiricism that has points of resemblance to the *"Dreck Apotheke"* of the ancients. For our present purposes what seems to me most noteworthy are, first, the persistence of such practices in medicine proper down to comparatively recent times, and, second, the traces of them that still remain in folk medicine.

But a glance at the old practices for the sake of background. Both Hippocrates and Galen prescribed "excrementitious material" for diseases without number. If such critics as Aristophanes laughed at "excrement-eaters," a concession might result, and human urine or dung be replaced by those of animals in weird variety: camels, goats, wolves, hens, wild boars, asses, calves, lynxes, crocodiles, bullocks, geese, oxen, sheep—each, at one time or another, as though it were specific. Pliny, Bourke says, prescribed as follows:

> Camel's dung, reduced into ashes, and incorporat with oile, doth curle and frizzle the hair of the head, and taken in drinks, as much as a man may comprehend with three fingers, cureth the dysenterie; so doth it also the falling sickness. Camel's piss, they say, is passing good for Fullers to scour their cloth withall; and the same healeth any running sores which be bathed therein. It is well known that the barbarous nations keep this stale of theirs until it be five years old, and then a draught thereof to the quantity of one hermine is a good laxative potion.

Galen is said to have disapproved of the pharmaceutical use of human feces because of its abominable smell. He also disagreed with Xenocrates, who recommended the internal and external use of sweat, urine, menstrual fluid, and ear-wax in medicine. Galen, Bourke tells us,

> thought that the internal employment at least of such disgusting curatives is of questionable expediency, especially when more agreeable remedies may be available. This objection would, of course, apply with special force to cities, although he admits that travelers, country people, and those suffering from poison, must use the first thing within reach.

Lucian, in his treatise on remedies for the cure of gout, says,

> And bones, and skin, and fat, and blood, and dung,
> Marrow, milk, urine, to the fight are brought.

The great Arabian physician of the eleventh century, Avicenna, shared the general enthusiasm of the time for stercoraceous remedies, but Averrhoës, his compatriot of the next century, seems not to have used them.

These practices flourished in Europe during the Renaissance and evidently were used more and more widely up to the seventeenth and early eighteenth centuries, doubtless stimulated by the rise of fearful epidemic disease for which no real remedy was available. Bourke says:

One's own urine was drunk as a preservative from the plague. Bekherius [1660] says he knew of his own knowledge that it had been used with wonderful success between 1620 and 1639 for this purpose. [Urine was also recommended] as a drink for lues veneris [syphilis]. . . . To smell human ordure in the morning, fasting, protected from plague.

Such great names as Paracelsus (1493–1541), Van Helmont (1577–1644), and Boyle (1629–1691) are among those recommending remedies of this sort. With developments in chemistry a kind of sophistication is applied. Thus Michael Etmuller (1690) prescribes:

From the urine of a wine-drinking boy, "urine pueri (ann. 12) vinum biventis," distilled over human ordure, was made "spiritus urinae" of great value in the expulsion of calculi, although it stank abominably: "sed valde doetet." This was employed in the treatment of gout, calculi, and diseases of the bladder.

Such "spiritus urinae per distillationem" was supplemented by "spiritus urinae per putrefactionem," as follows:

The urine of a boy twelve years old who had been drinking wine, was placed in a receptacle, surrounded by horse-dung for forty days, allowed to putrefy, then decanted upon human ordure, and distilled in an alembic. . . . The resulting fluid was looked upon as a great "anodyne" for all sorts of pains, and given both internally and externally, as well as in scurvy, hypochondria, cachexy, yellow and black jaundice, calculi of the kidney and bladder, epilepsy, and mania.

By 1725, when a treatise called *Chylologia*, compiled from nearly seven hundred authors, was published in Dresden by Schurig, opposition to such practices was growing. Yet in 1730 John Quincy of London, in his *Complete English Dispensatory*, was still recommending salts distilled from the urine "of a sound young man, newly made" for rheumatism and arthritis; and indeed favorable reports by physicians continued to appear in the last half of the nineteenth century. Thus Bauer, in 1852, reported that urate of ammonia and guano (dung of sea fowl) were useful externally in tuberculosis, leprosy, and various obstinate skin diseases. Hastings (1862) reported on the sale of reptile excreta for the treatment of tuberculosis. As late as 1881, in the midst of the bacteriological era, the paper of Neal mentioned earlier reported:

In South America urine is a common vehicle for medicine, and the urine of little boys is spoken of highly as a stimulant in malignant small-pox. Among the Chinese and Malays of Batavia urine is very freely used. One of the worst cases of epistaxis [nosebleed] ceased after a pint of fresh urine was drunk, although it had for thirty-six hours or more resisted every form of European medicine.

According to a report in the *Transactions of the American Philosophical Society* in 1889, urine was used in folk medicine in the United States, Canada, and England in the early nineteenth century for chapped hands, to prevent cramp, to ward off "fits," and as a laxative. Bourke cites the report that among German peoples in Pennsylvania and Illinois,

> "The white, limy part of hen-manure was used for canker-sores of the mouth." . . . Lamb or sheep dung or "tea" made from it was used to cure measles, cow dung in poultices to treat diphtheria. "Tea made of sheep-cherries . . . is given for measles."

Some of these preparations had euphemistic names. "Sheep-nanny-tea" was made from sheep-dung; human excreta masqueraded under such names as "zibethum" and "oriental sulphur." "Sheep-nanny-tea" sweetened with sugar, according to a letter to Captain Bourke from Professor S. B. Evans (1888),

> was used in the belief that it was "singularly efficacious in bringing out the eruption" of measles in Iowa, and thought to have originated in Indiana and North Carolina.

The practice is said to have been common until twenty years before that time.

Of body products other than feces and urine there is a miscellany of relics of sympathetic magic. According to Pliny,

> The first hair cut from an infant's head will modify attacks of gout. . . . The hair of a man torn down from the cross is good for quartan fevers . . . The smell of a woman's hair, burnt, will drive away serpents, and hysterical suffocations, it is said, may be dispelled thereby. The ashes of a woman's hair, burnt in an earthen vessel, will cure eruptions and porrigo [ringworm] of the eyes . . . warts and ulcers upon infants . . . wounds upon the head . . . corrosive ulcers . . . inflammatory tumors and gout . . . erysipelas and hemorrhages, and itching pimples.

Hair chopped fine or as ashes, applied externally as a salve, was also recommended, as one might guess, for baldness; but by more obscure reasoning for yellow jaundice as well, for luxation of the joints, and for hemorrhage from wounds. In Devonshire and Scotland, whooping cough could be transferred magically from a child to a dog by appropriate use of hair. Ear-wax (cerumen) and other "sordes," including the fatty matter of wool, were recommended for a variety of afflictions. Van Helmont said in 1662 that cerumen was "a great comfort in the pricking of the sinews." Human milk mixed with zinc sulfate was good for red eyes and obstinate hiccough. A butter made from human milk was recommended for colic as well as for eye difficulties in children. Pliny said:

> If a person is rubbed at the same time with the milk of both mother and daughter, he will be proof for all the rest of his life against all affections of the eyes. . . . Mixed with the urine of a youth who has not yet arrived at puberty, it removes ringing in the ears.

In a work on remedies derived from the human body, written by Samuel Augustus Flemming (1738), Bourke found the following which he evidently translates from the Latin:

> If the perspiration of a fever-stricken patient was mixed with dough, baked into bread, and given to a dog, the dog would catch the fever, and the man recover. It [sweat] was efficacious in driving away scrofulous wens, and in rendering philters abortive.

This is followed by the item on love-philters given in the last chapter.

Similar magical devices were described by others. Etmuller (1690) related that pieces of skin, excrement, or "anything else intimately connected" with a person, who may or may not have been the patient, were used to bring about cure by watering seeded soil or a growing plant, grafting on to a tree, or feeding to animals. Detailed instructions for these procedures were given. For instance:

> Take a sufficient quantity of ordure of a healthy man, and make it into a poultice with human urine, to which add sweat gathered from the body with a sponge; place this in a clean place in the shade until it dries, and when needed for use, moisten with human blood.

Fromman (1675) spoke of transferring or "transplanting" a disease by the method of "sympathy" or "magnetic transference," for instance by making a mixture of toe- and fingernails with urine, tying it in bags, and feeding it to chickens, or "throwing these in a road to be untied by some curious person who would catch the disease." Bourke suggests that Huckleberry Finn's cure for warts,

> Barley-corn, barley-corn, Injun meal shorts,
> Spunk water, spunk water, swaller these warts,

is a "distorted survival" of such a sympathetic cure.

PART III

For Better and Worse

12

The Great Scatologists

. . . as a certain kind of humour depends upon being able to
speak without self-consciousness of the parts and functions of
the body, so with the advent of decency literature lost the use of
one of its limbs. It lost its power to create the Wife of Bath,
Juliet's nurse, and their recognizable though already colourless
relation, Moll Flanders. Sterne, from fear of coarseness, is forced
into indecency. He must be witty, not humorous. He must hint
instead of speaking outright. Nor can we believe, with Mr.
Joyce's *Ulysses* before us, that laughter of the old kind will ever
be heard again.

—Virginia Woolf, *The Pastons and Chaucer*

This is a good place, I think, to stop for a moment and
look back. We have come to something like high ground. Not that
the road ahead is all downhill, or that we can be sure there will be
no more rough terrain. But the worst is probably behind us; the going
ought to be easier from here on.

What about these microbes we have on us, the places they live
in, especially the things they pick out to live on? How does it all
look—prejudices and all—as we see it from here?

A verse I remember from my youth, called "an ode on the antiquity
of microbes," tells us that

Adam
Had 'em.

Which puts it succinctly, if metaphorically. We abandon all pretensions to the contrary. They are all over our skin, burrowing under
what we see as the surface. They nestle in every fold and crevice.

147

They penetrate the nose and the moist urinary-sexual orifices, but not very deeply. The lungs, bladder, and uterus normally have no microbes. The eye has very few. But the alimentary tube from inside the lips all the way down is pretty densely settled; and down at the nether end of it are the largest numbers of all. We are not born with microbes. They come to us from outside, mainly from other people. It looks as though we get them through the same intimate loving contacts that we need to grow on, that we could least afford to do without.

Microbes are associated with disease, and so, partly for good reasons, they have come to be symbols of horror and loathing. But with little reason—how much, is one of the things we have yet to see—the microbes in our nooks and crannies fall under the common ban, so much so that even microbiologists prefer not to think of them. We have traced some of the history of these feelings, and found their roots deeper than our knowledge of microbes. Leeuwenhoek, whose discovery of microbes in general included those that live on us, hardly thought of being prejudiced against them, but he saw and recorded such prejudice in others. We have explored some of the trails over which these prejudices originated, how they grew out of the mysteries of fertility and sex, out of superstitions and notions of magic. We saw how such prejudice may have come to merge with feelings of contempt and aversion for such lowly things as parasites. But much earlier a groundwork of fear and taboo had been laid upon the places and things in ourselves in which we later found microbes. This turning away or rejection has been part of our culture for so long that we seem to have been born with it; but we were not, no more than we were born with the microbes themselves. What we now, without thought or hesitation, think of as the most repulsive or loathsome of materials, man once cherished for real virtues and idolized for imagined ones. Customs we can hardly bring ourselves to mention, or mention only after an appropriate disclaimer, have been practiced shamelessly until recently—perhaps they still are!—by people who seem, if we can only look at them with sympathy, not very different from ourselves.

The microbic population on any one of us is undeniably large, numbering vastly more than all the people on earth. Yet they could all be packed into something hardly bigger than an ordinary soup can; and the ones everywhere except in our guts—including our whole expanse of skin—could hide in the bottom of a thimble. Altogether, guts and all, they make up less than half of one per cent of our body weight. Think of them, nevertheless, as many rather than few; we know that if we could get rid of them without sacrificing those cherished relationships with people that brought them to us, we would regret the loss. The microbes themselves are evidently indispensable to our healthy development and well-being. Animals we can make live without microbes, in the artificial world of the germ-free laboratory, turn out to be puny and deformed, an easy prey to almost any passing infection, with deficiencies and weaknesses yet to be counted.

And so we seem to have been living for ages with an accumulation of mistaken ideas. We have been fooling ourselves.

Through the centuries, as we let this burden of prejudicial nonsense grow on our backs, there have always been individual men suggesting that we could put it down. They have been artists rather than scientists, rather even than philosophers. As a scientist I would like to be able to report that art used means provided by science to free us of prejudice; but it is not so. Science is the fountain of knowledge— assured, verified knowledge—and knowledge makes us free. But with only intuitive assurance and without what science could accept as verification, artists—a few of them—seem to have known all along what scientists have not learned even now.

If we grant that our burden of prejudice grew on us as a legacy of a perverted Christianity, and multiplied like a great tumor during the first millennium and more after Christ, we need not go back before the Renaissance to find the particular artists we are looking for. It is an odd coincidence that profound events evoked by microbes which themselves were as yet hardly so much as imagined played a part in the awakening. The poet-musician Guillaume de Machaut and Boccaccio the storyteller both acknowledged their debt to bubonic plague. Chaucer must have been influenced by it, too; but we know that he learned verse forms from Machaut and borrowed tales from Boccaccio.

But there was more to it than that. It has been suggested, no doubt with the oversimplification that sometimes helps us grasp the complexities of history, that the Black Death of the mid-fourteenth century in Europe, by underscoring the fragility of life, released man's long-suppressed affirmative spirit. A few short years thereafter, those still alive had expressed their joy in living by making good the human losses sustained in the pandemic, and the Renaissance had started. Boccaccio's *Decameron* begins with a description of the plague in Florence that has become classic in the annals of epidemiology. Machaut had written of it, more briefly but with equal feeling, in his *Judgment of the King of Navarre*. And, speaking of microbes, the spirochete of syphilis—then seriously rampant and epidemic—joined forces with the plague bacillus in an unmistakable effect on Shakespeare's plays.

But, as I say, this is too simple. Great historical events, like prejudice and life itself, do not spring forth in a day. Explosions and mutations themselves emerge out of antecedent events. There is no lack of evidence that the affirmative spirit never disappeared during the Middle Ages, which is as much as to say merely that the spirit of man was never quenched. We have seen it persisting in folklore, and we know that unsuppressed merriment never died out at higher social levels as well. We know this, for example, from what remains of the songs of troubadours and minnesingers, from the *Novellino* upon which Boccaccio himself drew, the fabliaux, the *Carmina Burana*. Chaucer and Shakespeare used these materials. They were rivulets of the human spirit that had never dried up.

The stream burst its dam and tumbled forth in joyous profusion in François Rabelais. He was "the merriest mind that ever existed"; "*le vray grand esprit universel de ce monde*"; "the father of free thought." We have seen him disparaged, and he never lacked detractors. Even Montaigne thought him no more than amusing. Voltaire dismissed him as a buffoon. But his genius is not more in question than Shakespeare's. The sensible thing to do with both is what is now pretty well agreed upon for Shakespeare: to waste no energy in praise or blame but simply to understand, and so to enhance appreciation and enjoyment.

Rabelais was a physician. He was also a priest (he "died an excellent Catholic and an exemplary pastor"). And he was, as well, pagan, humanist, philologist, man of all learning, "a veritable Man of the Renaissance." He has been called obscene—to be sure!—but he will doubtless outlive the reproach. His language was evidently current in his time at all levels. He reported it with no foolish restraint, no shame, with Falstaffian, belly-shaking merriment—vulgar, if you insist, but in the proper root sense of the word. *Alle Menschen werden Brüder.*

He was a churchman, and he knew his enemies. Samuel Putnam says:

Rabelais hated Calvin and all the ratty tribe that came out of Geneva, like poison. It was as though he saw that this ugly thing being born was to be responsible for all the ugliness of our modern civilization, achieving a culmination in the little frame meeting-house on the American prairie, in Evolution Trials, Ku Klux Klans . . . etc., etc.

Among scatologists Rabelais was not the first but surely the most exuberant. Doubtless the very exuberance, like the explosive force of what we call dirty jokes, was a reaction to the rising repression. Chaucer, by comparison, is almost casual. Villon, only a generation or so before Rabelais, seems to have vented all his scatological force in one poem, the "Ballade for Fat Margot"; he was, after all, too deeply immersed in pain to laugh like the others. But we have decided, haven't we? that laughter is not sinful. Rabelais laughed. It is his glory.

We need little more from him than one outstanding example. Speaking of Gargantua, aged three to five, in what is thought of as a satire on the education of princes, he says in Book I:

> He pissed over his shoes, dunged in his shirt, wiped his nose on his sleeve, dropped snot in his soup and paddled around everywhere.

And two chapters later we come upon the classical story of Gargantua's invention of the rump-wiper, at about the end of his fifth year of age, as he tells it with boyish enthusiasm to his father, Grandgousier. A long list of materials falling to the hands of one who "paddled around everywhere" is enumerated, with findings as to the merits of each, including the abrasive effects of the gilt on some bright satin ear pieces, and the "exulcerating" action of a cat's claws. Paper appears in the list without emphasis; the versified testimony dismisses the product of the age as unhygienically non-absorbent. The verse, incidentally, which includes another item listed as such and a roundelay, must be among the earliest of scatological graffiti, unless archeologists have been hiding things from us.

The conclusion of Gargantua's investigation, as you probably remember, fixed on the neck of a goose, to which, in preference to asphodel, ambrosia, or nectar, the young scientist attributes the happiness of the heroes and demigods in the Elysian Fields.

Hard on the heels of Rabelais was his countryman Eyquem de Montaigne, who was twenty years old when Rabelais died in 1553, and whose sober essays seem as far removed from the exuberance of his predecessor as one can imagine. But merriment or sobriety is not the point, except incidentally. Montaigne in his way was as uninhibited as Rabelais. He seemed to set no limits to the range of his interest. Many of his ideas seemed to come from a kind of self-analysis, supported by his reading of classical, especially Roman, authors. Rabelais, on the other hand, as a doctor—although he seems to have read everything available in his time—was always out among the people in his laughing and storytelling.

Montaigne is as free as Rabelais in his treatment of both sex and scatology. He reverts to such subjects less often, perhaps, but often enough to leave no doubt that he thought of them with as little constraint as of anything else. The question suggests itself, why has the world elected to damn Rabelais, leaving Montaigne with nothing but respect? Two possible reasons both bear on my argument. The first is precisely that Rabelais took pleasure in his writing and clearly meant to amuse his readers, who, laughing, see him doing the same. But Montaigne seldom provokes so much as a smile as we read him, and we see his face as impassive. Secondly, Rabelais is always affirmative toward man, even when he seems to anticipate Voltaire's *Candide* and Swift's *Gulliver's Travels* in satirizing human foibles. But here and there Montaigne seems at least temporarily committed to abnegation. Naked man, he says, has "manifold imperfections" and reason to cover his nakedness with adornment taken from animals that are more beautiful than himself. In an effort to be scientifically or philosophically objective, he can find in love only "an insatiate thirst of enjoying a greedily desired subject"—a notion he seems to struggle with and resolve, many pages later, with Cicero's "All things are to be counted good that are done according to nature." The difference may again be only humor or the lack of it; but these doubts, even though resolved affirmatively, may have brought Montaigne through the scrutiny of Puritan and censor.

Essentially Montaigne, like Rabelais, was a healthy-minded Renaissance Frenchman, with virtues beyond those usually credited to him. He could speak with utmost frankness of excretion in all three physical states, of the foolish custom of storing one of them in a handkerchief, and of other sorts of foolishness I have mentioned before (at the head of Chapter 8). He gives us one of the early bibliographies of pornography. He mocks the law that caused "ancient statues to be gelded," suggests that knowledge of such things may be better than fantasy, and goes so far as to raise the question,

> What harm cause not those huge drafts or pictures which wanton youths with chalk or coale draw in each passage, wall, or stairs of our great house?

With a frankness rarely matched (I have heard that Benjamin Franklin said something similar), he advises a limping woman as a more perfect bedfellow than one unhandicapped, and attributes similar sentiments about "the crooked man" to the Queen of the Amazons. He has homely advice for all defecators, among whom he boldly includes kings, philosophers, and ladies, not to become dependent on a particular "commodious *Ajax*," and to be what we call "regular" in our excretory practice, but not too compulsively:

> ... myself never miss to call one upon another at our appointment, which is as soon as I get out of my bed except some urgent business or violent sickness trouble me.

And finally, working his way toward Cicero's rule, he tells us that he has "willingly seconded and given myself over to those appetites that pressed me." He argues against trying to cure one evil with another: "I hate those remedies that importune more than the sickness." A mature philosophy is compressed in these words:

> Since we are ever in danger to misdo, let us rather hazard ourselves to follow pleasure. Most men do contrary, and think nothing profitable that is not painful; facility is by them suspected.

I have made no attempt to be complete in this listing of the literature of scatology: such an undertaking would require a book in itself. The High Renaissance has much more. As a passing example: The *Heptameron* of Margaret, Queen of Navarre (1559), modeled on the much more vigorous earlier *Decameron* of Boccaccio, contains an account of two lovers who meet in a garderobe, a "nasty adventure" that befell a lady in a dark privy, and more to similar effect. Miners for Freud's gold will have no difficulty finding more. Let us move on to Shakespeare.

One can, of course, find anything one looks for in Shakespeare, as in the Bible. Shakespeare holds up the mirror not only to nature but to the individual reader, so that what we see in him may be the reflection of ourselves, or perhaps of what we would like to be. I see him as the very incarnation of Renaissance affirmation. He explores the texture and substance of man with unbounded joy, or with loving sorrow, or with compassionate pain. He understood, as men generally seemed to do before the Puritan blight came upon them, how dominant a force in our lives sex is—an understanding subsequently hidden from polite eyes until Freud exposed it again. He matched the clear view of Rabelais and Montaigne, and capped it with poetic vision. He came nearer to understanding us than anyone else I know of, with a special clarity that seems possible only for poets.

Now that the scholar Eric Partridge has given us back the meaning of all his words, we can understand him again almost as well as the groundlings and aristocrats, both, are said to have understood him in his day. He has substance for every level of experience. Much of it is accessible to children; and some of this, unexpurgated and with suitable glossaries based on Partridge, might be considered for their education, preceding selections from Freud. Children are likely to know what he is about and to make ludicrous the efforts of censors to protect them against him. At a matinee performance of *The Comedy of Errors* I recently heard schoolchildren laugh with pure joy at the Syracusan Dromio's uncut geography (III, 2). They evidently understood enough of it and found nothing to blush at. Only grownups are offended by such words; and it is inconceivable to me that anyone could be harmed by them.

Scatology finds its proper place in Shakespeare—meaning that, compared with sex, it is inconspicuous but not insignificant. Partridge has done the same service for it, including it within the scope of his "bawdy." His book is in paperback and easy to come by.

When Shakespeare died we lost much more than England's greatest poet. What happened is usually dated with the death of James I nine years later, in 1625. Not that these momentous events are easy to trace in the history books, which delight in covering important things with a protective screen of political and military trivia. Certainly the Puritans, who came to power soon afterward, are the key to the story. But causes and mechanisms are much less clear than consequences. We know what happened, but not precisely how or why. The times are a tangle of religious conflict, war at home and abroad, ferment in science and philosophy; it is the time of Milton and Purcell. Historians dwell lovingly on every strand, it seems, except the one I would like to trace. During this interval the English-speaking people surrendered a freedom—the freedom of Elizabethan speech. We are told that they gained other freedoms; but I suspect that the one they threw away was more precious than what they got, that they may have lost the most beautiful and possibly the most potent instrument they ever held in their hands. Science has given us all manner of other instruments since, and I am the last to disparage them. But the loss remains.

Agree with me or not on the importance of what happened—nobody doubts that it happened. It is not for nothing that we speak of Elizabethan monosyllables. Nor is it denied that the critical events happened in the interval between the accession of the puny Charles I in 1625 and the Restoration of his son in 1660. But there was no single determining event. Friedrich Heer speaks of

> the "merry old England" of fairies and magicians, of poets and showmen, buffoons and drink-loving sages, which shortly after Shakespeare's death went under in the storm of sermons unleashed by puritan and nonconformist ministers, and disappeared forever in the revolution.

The gradual character of the change is suggested by events during Shakespeare's last period that seemed to foretell it, and others that happened only after 1660. These latter are reflected in the history of censorship in England. While matters "heretical, seditious and schismatical" had been banned in one way or another since soon after the invention of printing, it was not until 1662 that the first Licensing Act attempted also to suppress "offensive books and pamphlets." Even so, chapbooks and similar light literature regarded as very coarse, and madrigals and catches with bawdy verses, seem not to have been interfered with under the act. The kind of censorship we know today came much later. When the 1662 law expired in 1695, it was not renewed. The fact is taken as evidence for the vaunted British "liberty of the press"; but there seems rather to have been an abdication of that liberty. I see signs here of what we know as commonplace from direct experience today, that the so-called facts of history can be turned inside out to look exactly the reverse of what they were.

The change in English speech that took place during this time may be keyed to the closing of the theater by Puritan edict between 1642 and 1660, when Charles II opened it under royal patronage immediately upon his accession to the throne. This sharp break makes the change easier to see and points up what happened in the meantime.

In the Restoration theater, elaborate and movable scenery and stage machinery replaced the simple settings of Elizabethan days. Women now took female roles in place of boys. The plays concentrated on comedy, and the whole enterprise was directed toward the upper classes alone. Partly hidden under these changes, many of which have persisted into our own time, was a substitution of visual realism for poetic suggestion, an emphasis on form at the expense of content, an abandonment of wholesome bawdry in favor of mannered licentiousness. Some of these novelties had been imported from France by Charles and his courtiers; but on English soil the transplant had a different, and certainly a less robust, growth than in the Augustan age in France of Corneille, Racine, and Molière.

But the change was not so simple. The process had been anticipated in the staging of Shakespeare's last plays, as in the use of stage machinery in *Cymbeline* and *The Tempest*, and the elaborate setting for *Henry VIII*. It had continued in the later Jacobean theater, including the later plays of Ben Jonson and the stagecraft of Inigo Jones. Nor need it be suggested that the Restoration theater was tawdry or trivial. As new plays were written, the Elizabethan dramas were, in fact, revived, Shakespeare's among them; but they were abridged and altered. *Antony and Cleopatra* was turned by Dryden into *All for Love*, said as recently as 1933 to have been an improvement, even though it "allowed much of the human essence to escape." A *Midsummer Night's Dream*, of all Shakespeare's plays the one usually thought to be most acceptable for children in our day, reappeared as *The Fairy Queen*, embellished, to be sure, with the immortal music of Henry Purcell. Perhaps for the first time, as we gather from some of Pepy's remarks, the playgoer was more concerned with the performer than with the play. We know from experiments in the theater in our own day that when Shakespeare's and other Elizabethan plays are mounted in the simple intimate settings of their own time, and embellished only with the means they used—the stirring, gay or sad music, lavish colorful costumes, deft stage business, and, possibly most important (some of the rest can sometimes be omitted without great loss, as in the Gielgud–Burton *Hamlet*), a sustained fast pace, but not too fast to let the words come through!—the result is amazingly successful. Today such experiments are reactions to the literalism of the movies and the infantilism of television. In the theater the illusion can be strongest when distraction is minimized, when everything is subordinated and made complementary to the magic of the words. We cannot quite recapture the verbal responses of Elizabethan audiences, the quick appreciation of both lords and groundlings. Theirs was a time of widespread illiteracy. They depended on words spoken, repeated, intoned, memorized. Their ears and their minds were tuned accordingly. But we can glimpse the magnificence of Elizabethan speech and mourn our loss.

The change began while Shakespeare was still alive, and was not complete in Restoration times. It never really became complete. Dryden himself, a key figure in the literature of his time, did not entirely abandon the little Elizabethan words, as we have seen in his translations of Juvenal, to which he lent some of the color of his age. Like other manifestations of the time, Restoration drama reacted to Puritanism and reflected a dissolute, cynical court. Class distinctions were accentuated. The linguistic revolution that introduced the Renaissance, when the classical languages of scholarship and the Church were replaced by the vernacular, the languages of the people, was being undone. This process continued into Victorian times, and has begun to move the other way only in the twentieth century.

A symptom of the verbal disease of the eighteenth century is the peculiar swearing of Acres in Sheridan's *Rivals*. Gods or other perilous sacred names, or word-symbols extending from sacred to profane to only forbidden, are now pushed aside altogether in favor of merely amusing makeshifts: "odds frogs and tambours!" "odds flints, pans and triggers!" It is *genteel*, he says, the *"oath referential, or sentimental swearing."* Maybe it is an oblique way of saying, this is what we have been reduced to.

An important witness of the Restoration was Samuel Pepys, perhaps the more so because his diary was intended for no eyes but his own. Noted for his frankness, he nevertheless regularly hid behind euphemism and circumlocution. He is ashamed of his love letters to his wife, wresting them from her in a quarrel and burning them for fear "of so much disgrace to me and dishonour, if it should have been found by anybody." In another of his recurrent domestic squabbles he is vexed that his wife abused him with so colorful a word as "pricklouse" (a scurrilous name for a tailor); and he complains a little later that

> my wife did say something that caused me to oppose her . . .
> she used the word "devil" which vexed me, and among other
> things I said I would not have her to use that word . . .

It is typical of Pepys and of his time that he speaks of reading a "most bawdy, lewd book," dipping into it repeatedly with evidently mixed pleasure and guilt, and eventually burning it "so that it might not be among my books to my shame." This is a long way from Montaigne.

Pepys was not a skillful writer, and the fact may make his testimony the more valuable in being less colored with art. His famous account of the plague in 1665, of which he was an eyewitness, is worth comparing with that of Defoe, who was five years old then, and who is thought to have transposed to London what he saw in Marseilles in 1720. Pepys' account is interspersed with comments on the Dutch war and his own petty private affairs, and is colorless when placed beside Defoe's vivid reporting. Defoe himself came, in the generation after Pepys, to be one of several authors who dealt in a new way with sexual freedom, in *Moll Flanders* and *Roxana*. (And, incidentally, there are hints throughout this survey that the prevalence of scatology, which is easier to suppress than erotica, may be a more sensitive index of verbal freedom.) Pepys testifies to the persistence of erotic literature in his time. There seems to have been plenty of it, a plaything for ladies and gentlemen, dressed in the fancy verbiage they affected, to match their brocades, ribbons, and powdered wigs.

The early eighteenth century in England saw a revival of a more broadly based realism of expression by writers like Fielding and Sterne, of whom I have already spoken and need say no more. John Gay's *Beggar's Opera* supplied "popular" music as an antidote to Handel's Italian finery, and both subject-matter and language appropriate to its avowedly lower-class theme. But the towering figure of this period is surely Jonathan Swift.

Swift is widely regarded as the most scatological of all serious writers, perhaps not only because he reverts to such matters so often but because, in addition, he is seldom or never erotic: "sometimes coarse, but never lewd," as the critic Sir Paul Harvey put it, presumably intending extenuation. Norman O. Brown, in his psychoanalytic (and extraordinarily scatological) study of history *Life against Death*, devotes a long chapter to Swift and his critics and concludes that Swift anticipated Freud's theories of anality and sublimation. The significance of suggestions that Swift was sexually inadequate may be mentioned but put aside; the scatology, in my non-Freudian view, can be displayed, and even judged, on its separate merits. Especially in relation to Swift, Brown is able to illuminate both the penetrating insights of Freud and the diagnostic absurdities of some later psycho-analysts, professional and amateur: I profess to be neither.

The most relevant of Swift's works are, in chronological sequence, *A Tale of a Tub* and its sequels (1704), *Gulliver's Travels* (1726), and the poems "The Progress of Beauty" (1720), "The Lady's Dressing Room" (1730), and the pair "Cassinus and Peter" and "Strephon and Chloe" (both 1731). From an over-all view of both the imagery and the frank scatology in these works, my impression is one of an attempt to restore excremental facts to their rightful place in human life. That place is in fact no less than the one it occupies in the totality of Swift's work. We ought to compare Swift with the life he describes rather than with the myopic standard set by other churchmen and writers. He has been found shocking, and indicted accordingly, in proportion as he has pointed out that excrement exists and cannot be abolished either by sublimation or by romantic illusion. His Chloe,

> By Nature form'd with nicest care,
> And faultless to a single hair

nevertheless, having consumed twelve cups of tea on her wedding night,

> Steals out her hand, by nature led,
> And brings a vessel into bed;
> Fair utensil, as smooth and white
> As Chloe's skin, almost as bright.
>   Strephon, who heard the fuming rill,
> As from a mossy cliff distil,
> Cried out, Ye Gods! what sound is this?
> Can Chloe, heavenly Chloe, ——— [*sic*]?

I am again as much impressed with the observer as with what is observed. The edition of Swift's works from which I have taken the poems (P. O'Shea, New York, 1863, Volume 1), gives all of "Cassinus and Peter" except its last two lines, without which the whole poem is meaningless! But an obliging earlier scholar had found the missing lines and written them in by hand; and since they agree with Brown's version (but supply the word of which he gives, at best, only the first two letters), I take them to be correct. Again, a romantic young man, one of

> Two college sophs of Cambridge growth,
> Both special wits, and lovers both,

is sick to the point of death over his beloved Celia, not for any of
the expected reasons; she is not dead; she hasn't "played the whore";
she doesn't have "the small or greater pox"; no, she is, instead, guilty of

> A crime, that shocks all human kind;
> A deed unknown to female race,
> At which the sun should hide his face:

which emerges (handwritten in my copy, as I say) after the suspense
of some fifty more lines as

> Nor wonder why I lost my wits;
> Oh! *Caelia, Caelia, Caelia* shits.

Swift demonstrated that the forces that distorted English thought
in the seventeenth century damaged more than they destroyed. His
vision is again that of Rabelais and Montaigne, but his method is
satirical instead of jocular or straightforward, and his language is not
as free as theirs. He has also acquired traces of prejudice; but these
I am inclined to forgive as the scars of the battle. It seems unlikely
that he intended *Gulliver's Travels* as a children's book, but rather
that he used its method as a device to convey his overriding concern
with the problems of England as they appear in his other works. But
children, absorbed in his tale, can both overlook the deeper satirical
commentary and absorb the scatology without blushing, their prejudices having not yet hardened. Called pessimistic and misanthropic,
Swift seems to me no more negative than Shakespeare was. Unlike
Pepys, he shows no cynical amused detachment; he can come very
close to man—or woman—and reveal with frankness what he sees
as defects or foibles; but in doing so he never separates himself from
what he sees. His basic loving identity with man—in my judgment—
never wavers.

At his most mordantly satirical there is nothing in Swift that I can
call detachment or rejection, not to speak of the alienation of present-
day "sick" literature. Man is too often a Yahoo, but he can be a
Houyhnhnm. Seen close, in Brobdingnag, there is ugliness to compare
with the delicate charm of Lilliput, where it is Gulliver himself who is
gross. This is the notion of ugliness at close range that appears in "The
Progress of Beauty," which peels the paint off a beautiful woman, and
most strikingly in "The Lady's Dressing Room," an inventory after
five hours of make-up in the days before "total cleanness" had been
recommended by science and facilitated by technology. The filth

under the paint in 1730 was probably not at all exaggerated as Swift described it: unlovely at best and the worse for being denied.

Gulliver's Travels, a product of Swift's maturity and the most accessible of his works, is also most revealing for my purpose in that it can be understood and accepted by children. The Lilliputian bonds released on one side, Gulliver turns and makes water, "very plentifully . . . , to the great astonishment of the People." The word is "astonishment" with no moral judgment. The Lilliputian reaction is merely quantitative. But a little later Gulliver speaks of both "Urgency and Shame" as he gives details of his first defecation in the new country. The result is an extraordinary story, balanced, mature. If there is anything to compare with it in English literature, I have not seen it.

Later Gulliver, being ordered by the Emperor to stand "like a *Colossus*" and let the troops march under him, offers a touch of pure Rabelais. Although the troops had been ordered to "observe the strictest Decency," some of the younger officers could not help

> turning up their Eyes as they passed under me. And, to confess the Truth, my Breeches were at that Time in so ill a Condition, that they afforded some Opportunities for Laughter and Admiration.

Still later we have the famous passage in which Gulliver puts out the great fire. Again there is a precisely balanced rightness in the description, helped with a show of deference to the outraged feelings of the Empress.

As I mentioned, Gulliver in Brobdingnag is impressed with a sense of the ugliness of man seen very close. One of the incidents is an eating scene that might have rivaled that in the *Tom Jones* movie, but Swift's microscopic vision is faulty. After several similar allusions he makes a casual reference to a microscope, and it is evidently on the basis of his actual observations through the crude toys of his day that these notions arose. Pepys mentions having bought such a microscope. It could have magnified little more than ten diameters. Swift seems not to have heard of Leeuwenhoek; but his idea might not have been affected by a look through the Dutchman's vastly more powerful lenses. His aversion to magnified images is enough to expose Gulliver's medical pretensions as fraudulent. The imposture emerges also in his failure to see disease, or to be sensitive to it as a physician could not avoid being. Gulliver speaks in one place of seeing, in

Brobdingnag, a woman with cancer of the breast, and his description is surely that of a person who might have peered at a mounted specimen through a crude microscope, and with a layman's preconceptions turned away in revulsion, much as Leeuwenhoek's "gentlewomen" turned away from the "little eels in vinegar." These are not the reactions of a person who professes to understand and treat disease, even though it is an eighteenth-century understanding.

But if the response is not medical it is still that of a healthy person who merely shared some of the prejudices of his day, as we share those of our own. In a similar context, Gulliver is disgusted by the smell that emanates from the breasts of young Brobdingnagian women. He recalls that a Lilliputian had complained of Gulliver's smell. He seems to assume that the smell objected to in sweat has nothing but a quantitative basis, although he gives evidence of suspecting the contrary: the person of his nurse Glumdalclitch was "as sweet as [that] of any Lady in England." As we know, the smell of sweat we think of as offensive is that of sweat decomposed by microbes. It is not the amount of sweat but the change in it on standing that we smell. It is also true, however, that the products of decomposition, or their relative predominance, evidently differ with diet. We object to the smell of sweat of unfamiliar people. It has been said that Asians, who eat a largely vegetarian diet, object to the smell of Europeans and Americans, who eat more meat. Gulliver restores his usual balance by observing that the natural smell of the young women of Brobdingnag was much more supportable than was the odor of their perfume, under which he immediately swooned away.

That Swift takes this whole matter of smell in moderation is apparent in the instance of Gulliver's defecation in Lilliput, when his excreta had been carted away in wheelbarrows by servants: the matter of smell is not emphasized. And later, when he is set down in a pasture in Brobdingnag and, trying to jump over an island of cowdung, he lands in the middle of it up to his knees, he speaks of being "filthily bemired," but the emphasis falls on the mirth at his expense the incident provoked.

As famous as the natural-fire-hose incident at Lilliput is the dung-scientist at the Academy of Lagado, who was working on

> an Operation to reduce human Excrement to its original Food, by separating the several Parts, removing the Tincture which it receives from the Gall, making the Odour exhale, and scumming off the Saliva.

I take this to belong in the same class as the degradation of the Yahoos in the last of the voyages. It is neither antiscientific nor antihuman, but an effectively droll satire on human error, something that man as scientist and simply as man recognizes, tries to minimize, never entirely overcomes. The essence of Swift is that he could treat such things with a deft touch, neither timid nor bold, with just the blend of cheerfulness and aversion his culture made unavoidable.

Passing from Swift to Tobias Smollett, we arrive at a scatological landmark. His *Humphrey Clinker*, written the year he died, 1771, seems best remembered for its lurid descriptions of matters fecal. It happens, perhaps coincidentally, that this is also the time of Goldsmith, whose *Letters* had been written a few years earlier. Both Smollett and Goldsmith had been trained as physicians. And, as we saw before, at this time *Dreck Apotheke* was still flourishing, although beginning to decline. These circumstances bear on the scatology in *Humphrey Clinker*, and on the way it differs from that in Gulliver. The details now are authentically medical, as is the language, which has moved so far from Elizabethan as to be heavily larded with words of Greek and Latin origin, and in fact with these and other languages themselves. The general tone is still entirely affirmative but has moved in the direction of a lusty exuberance like that of first-year medical students in the dissecting room, possibly a bit strong for unaccustomed stomachs. Witness an account of comparative reactions to "stink" in response to an opinion that the sulfurous fumes from a hot spring may be harmful. The countering opinion, put into the mouth of a doctor, amounts to the implication that "stercorous effluvia" are far more healthful than fresh air. There is also a description of the practice then current of discharging the contents of close-stools through upstairs windows, previously noted; and details concerning administration to "a fat-headed justice of the peace, called Frogmore," who had overeaten, of an emetic and an enema at the same time, with consequences in full.

Except for these incidents, which make up only a tiny part of *Humphrey Clinker*, the story is almost Dickensian in its human feeling and forecasts the Victorian novel in its propriety. Dickens is said to have been indebted to Smollett, but not, I think, to his scatology. For a century or more after Smollett, scatology, at least at the higher literary level, seems to have been in decline. Its submergence is in contrast to erotica, which continued to flourish in England with a succession of celebrated exponents, from Samuel Richardson, Boswell,

Burns, Byron, and the painters Hogarth and Rowlandson (with more than a trace of scatology in the last three, as we have seen!) in the seventeenth and eighteenth centuries, through such persons as Swinburne, Rossetti, Wilde, and Beardsley in the nineteenth century, and down to modern times with D. H. Lawrence, Frank Harris, and, of course, James Joyce. The frankly erotic work by John Cleland, *Fanny Hill*, circulated widely in the second half of the eighteenth century; while by 1885 the comparable *My Secret Life*, recently reissued openly, had to be anonymous and limited to a privately printed (eleven-volume) edition, as a gentlemanly commentary on Victorian morals.

With Joyce, in *Ulysses*, scatology emerged again, bringing back Elizabethan language. Yet the form of this work is such as to make it accessible only to the scholarly. Today it can be compared with Shakespeare, but only because we no longer speak the Elizabethan language. Shakespeare was not obscure in his own time, and it is obvious that Joyce did not mean to speak to everybody, as Shakespeare did. Nevertheless, *Ulysses*, above its intrinsic worth, served, once it cleared the hurdles of censorship, to reopen the whole field.

Not that it is wide open even today, or there might be no purpose in this book. We have, on the one hand, the easily available work of Mailer, Kerouac, LeRoi Jones, William Burroughs, and Lars Görling—to mention only a few—and we have the memory of Lenny Bruce, partly evocable in print. To me some of this is healthy and some is not, but the opinion is a personal one; I recommend that you judge for yourself, if you haven't done so already. Scatology also emerges—often surprisingly—in so glossy a periodical as *Ramparts* and in one so scholarly as the *New York Review of Books*. But none of this, if I judge correctly, reaches more than a limited audience, still, in fact, largely confined on class lines to the smallish group in which the habit of steady serious reading has not been displaced by television and comic strips. Those media, in common with the metropolitan press and the large-circulation magazines—the reading-matter of the people at large—remain untainted. I am grateful for fairly free mention now of venereal disease and contraception, subjects forbidden until recently, as well as other aspects of sex.

A more uninhibited scatology (and sex too!), evidently reaching a wide, intelligent, and vigorous audience, lurks in the so-called underground press, to be found in great variety and without special problems of access in campus bookstores, among other places. Some of this material betrays its lack of ease by protesting too much, shouting

too loud. It outrages delicate ears, and the offending papers suffer
continual persecution by the forces of law and order. But I find it
one of the more promising manifestations of today's youth. In rejecting
many of the values of their elders, including our limitations of speech,
they are on a road away from hypocrisy. Rabelais would have approved of them.

And Montaigne would have been keenly interested if he could have
seen this new phenomenon. He might have given an essay to it.
Curious human behavior was grist for his mill, and he was not inclined
to pass judgment. I think this quiet student of man came as close as
he ever did to showing delight when he spoke of a posterior trumpeter
whom Juan Luis Vives, Spanish contemporary of Rabelais,

> endeareth by the example of another of his days, who could let
> tunable and organized ones, following the tune of any voice pro-
> pounded into his ears, inferreth the pure obedience of that
> member, than which none is commonly more indiscreet and
> tumultuous. Myself know one so skittish and mutinous that these
> forty years kept his master in such awe that will he, or nil he,
> he will with a continuous breath, constant and unintermitted
> custom, break wind at his pleasure, and so bring him to his grave.

Irreverent, and no doubt exaggerated. But nothing said of man could
be more accepting; hence nothing could be more wholesome.

Here is a physician, Dr. F. Avery Jones, at the Central Middlesex
Hospital in London, who inserts a little paper into the journal *Practi-
tioner* for March 1967, entitled "Burbulence: a Fresh Look at Flatulent
Dyspepsia." Irreverent again, in the manner of Dr. Rabelais and Dr.
Goldsmith and Dr. Smollett. Dr. Jones begins by quoting from Ogden
Nash, to whom he credits the key word in his title:

> How do I feel today? I feel as unfit as an unfiddle.
> And it is the result of a certain turbulence in the mind
>   and a certain burbulence in the middle.

Dr. Jones proceeds to suggest, airily, I think, but not windily, that
Mr. Nash's word

> may be used to cover various windy syndromes: aerophagy, burp-
> ing, belching, bloating, flatulence, borborygmus, wind, distension
> and flatus.

We are warned by the title of the paper that its author does not
intend to delve below the stomach; but we can be irreverent too, and

extend his words past his limit. Aerophagy, he says, is swallowing air, which enters the esophagus and is belched on expiration. But properly speaking, "burping" is esophageal—the air escapes before it goes very far down—while "belching" is from the stomach. "Eructation" would include both. "Borborygmus" is "a noisy, windy abdomen," or the rumbling sound of gas being pushed along, either in the stomach or in the intestine. "Distension" (more commonly "distention") is "a sensation," although air or gas may actually be trapped in the stomach, or again, farther down. "Wind" (undefined by Dr. Jones) is usually associated with the verb "to break," and the consequence is sometimes called "flatus"; but this is merely a Latin word for gas or air and implies neither location nor direction. Another word, "crepitus," really referring to a crackling sound, is more likely to be used to mean *posterior flatus.* There is an apparent semantic lapse here; and it seems to me that instead of beating around the bush we ought to return to the little old English word "fart," which is more exact than anything else we try to substitute for it, and has the added virtue of serving equally well as noun and as verb. Dr. Jones does not use it; but he was concerned, as we know, with higher things.

Getting back to the good doctor, we find him speaking of a pathological fermentation in the stomach that may generate the gas methane in quantities "sufficient to give rise to an explosion if the patient eructates while lighting a cigarette." This is fortunately rare; but the comparable microbic effect in the lower intestine also generates inflammable and explosive gases, namely hydrogen sulfide and hydrogen itself. This happens much more commonly; and one can imagine that smoking might never have become popular if the evolution of mammals had not placed the anus at a good distance from the mouth.

13

Trouble

Cette notion d'une maladie causée par les saprophytes, inoffensifs par eux-memes quand l'enfant est bien portant et devenant subitement nuisible est devenue absolument classique.

—H. Tissier, Thesis, Paris Faculty of Medicine, 1900

(This idea of a disease caused by saprophytes, themselves harmless when the child is in good health, and suddenly turning noxious, has become quite classical.)

We have an incurable habit of judging other living things by ourselves. We attribute motives like our own to them and put values on them depending on whether and how much they do us good or harm us. The analogy I have implied in the title of this book, comparing man with a planet and our microbes with ourselves, is beginning to break down. Meaning to counteract the prejudice against microbes we are schooled in, I have tried to suggest that our microbes are as much a part of us as we are of earth. Now that I come to speak of the damage the microbes do to us, my partisan zeal suggests giving the analogy up. I am tempted to say, now, that of course such analogy is useful so long as it helps us to understand, but we mustn't push it too far. Hoist with mine own petard, I must either admit that where there is love there is also hate, or that I didn't really mean it to begin with. It looks as though a scientist has no business dabbling in metaphor.

I intend to abandon the analogy, but as a gesture in parting with it I can't resist reminding you that if you cling to it and attribute motives to the acts of microbes we are about to deal with, whether they be malice or only carelessness, you should think of the depredations man makes on earth. If *we* do *our* host any good at all, *we* value the benefit in terms of *our* comfort, convenience, and pleasure. From the host's viewpoint, if it could have one, or from the viewpoint of him, if there is anyone, who is at the other end of those signals from outer space, it is doubtful that our efforts would elicit admiration. We are in fact learning to deplore the way we destroy the earth's forests, kill its lakes, pock its surface, and poison its air with a ruthlessness distinctively human. If you persist in thinking of motives, take note that the plague bacillus and the spirochete of syphilis have better reasons for what they do to us than we have for what we do to earth.

Consider that point. The plague that comes to be called bubonic (there is also a pneumonic form, caused by the same bacillus, that is even worse) begins as a disease of wild rodents, such animals as marmots, gerbils, and ground squirrels. It is called "sylvatic"—*sylvan*, of the forest. It is a relatively benign disease of these animals, spread among them widely by their fleas. They tend to develop immunity to the bacillus and seldom die of plague except when a migrating group introduces the disease into a susceptible herd. It appears that when man began to establish himself on earth and to clear forests, he disturbed the rodents and started a cycle of events that brought plague to him. As the bacillus moves from one rodent population to another it may pass from rats living on the outskirts of human settlements to others living intimately with man. At each step the disease becomes more violent as it moves from more immune to more susceptible territory. When the house rats die in droves, their fleas, having no more rats to go to, jump to man and start the cycle of bubonic plague. The sequence calls for an assortment of appropriate conditions, including the kind of intimacy among people, rats, and fleas that was taken for granted during the Renaissance in Europe. The fleas jump from person to person, from dying to living, looking for warmth. A person with plague pneumonia, not yet too sick to move around, can start another cycle in which the bacillus, doing without fleas, goes directly from man to man via the cough. This is all biologically uneconomical, unecological. All life tends to perpetuate itself, but these cycles end in frustration. Rat, man, flea, and plague bacillus all die. The stable sylvatic arrangement, motives and human values aside, is as biological as a balanced aquarium: plague in man is a disease not only of man but of the whole system. Granted that none of it would happen without the plague bacillus, if blame is to be handed out, can man's meddling be excused?

I am for eliminating plague and syphilis with no sentimental nonsense. There is nothing tolerable about disease. But stop and think, just the same, how exquisite an example of parasitism syphilis is. The spirochete that causes it, cousin and near-twin of one of those in our gum clefts, has evolved a balanced arrangement almost as stable as sylvatic plague. It provides for itself a long, comfortable sojourn in man, who nearly always dies of something else, and a nice routine long before that happens for moving over to another host and starting again. The pattern has finesse. These operations of nature show a mutuality that man tends to overlook in his regard for his host, the earth, and that he can be brought to exercise toward other living things, even toward his own species, only in moments—or so it seems—when he forgets to be predatory and ruthless.

But—and this is good-by to analogy—the microbes, from one end of the scale to the other, are of course responding to physical and chemical forces in their environment with the equipment furnished by their genes. They have no nervous systems. They lack even the rudiments of a moral sense. The ones that adapt themselves to life on man tend to survive with man. Their evolution has been guided by survival factors that make for mutuality, for a long-term stable relationship in the individual host and plenty of easy ways of being transferred from one host to another. But on the other hand, their survival also requires that they withstand the host's active efforts to dislodge and destroy them. They are armed against these efforts and adapted to make use of what the host offers them as nourishment. Some damage to the host is unavoidable in this arrangement, and additional damage is potential in it.

Even in associations between different species apart from man and uncolored by prejudice, we now recognize that mutual damage is as much a part of the bargain as mutual benefit: we can't have one without the other. A classic example of symbiosis is that of a bacterium living in the root-nodule of a legume like peas or clover. The bacterium, a species of *Rhizobium*, using for energy carbohydrate made and supplied by the host plant, transforms atmospheric nitrogen gas into a combination that can be built by the plant into the amino acids of its protein structure. This is the mutually advantageous side of the partnership. But the interaction of the microbe in the root-nodule with the plant is now seen clearly as "infection," with evidence of disease-like alteration in both partners. It is a bargain struck and paid for. Indications of a similar balance are found also among the lichens, that group of curious pairs, fungi married to algae, so intimately blended that they were long considered as single species. Here, in opposition to the "idyllic" relationship formerly assumed, we are learning that the more independent alga serves as "host" to the more parasitic fungus in a relation of true infection. Damage is more apparent in the alga, but interaction cannot be absent; for what it gets, the fungus, we need have no doubt, must give in return.

It ought not to be surprising, then, that the life on man is not to be lightly dismissed as "harmless." We know it does us good, and we have to be prepared to see a price paid as our part of the bargain. The life is only harmless, or its effects on us are only entirely tolerable, within certain limiting conditions. When these limits are overstepped it becomes damaging, destructive, even fatal.

The point needs some emphasis. It was advanced in the early years of bacteriology by German and French scientists—as witness the quotation at the head of this chapter—but it tended to get lost. It got swallowed up in the rush to identify "pathogens" and control the serious infectious disease then raging; the problems we are coming to speak of often seem trivial, or, if serious, they seemed refractory to the early approach and were brushed aside. That early approach was astonishingly successful and productive, and it may have made the whole enterprise deceptively simple. Notions of "either-or" grew up: a microbe is either a pathogen or it is not. It can't be both harmful and harmless. But classification of anything biological is notoriously ambiguous. There is always a middle ground; it is not possible to avoid gradations and blurred boundaries. But the fact never stops being annoying. After a while an amendment was proposed. Certain microbes were spoken of as "opportunists." The name bespeaks analogy again—microbes sharing human vices. Anyway, it leaves the basic fallacy intact and provides no useful further meaning. All microbes, all living things, respond in some way to changes in their situation. All possible degrees and kinds of opportunity change harmless microbes to harmful ones. Plague in wild rodents may smolder until their fleas find a susceptible herd. Meningococci can live quietly in a human thoat and accidentally break through to the blood and brain to cause rapidly fatal meningitis. The typhoid bacillus, its work done, may nestle cozily in the gall bladder, leaking out in feces to contaminate food and water and cause typhoid fever in others. There are any number of further examples.

Of course the microbes of plague, syphilis, cerebrospinal meningitis, and typhoid fever are all "pathogens" in a sense that does not apply to the spirochetes of the mouth, the twiglike microbes of the skin, the bifid bacteria of the nursling's intestine, or the colon bacillus. The difference needs to be marked, but not too sharply or arbitrarily. It is not all or nothing; it is not even great or small; it is more a matter of obvious or subtle. But no simple distinction will be infallible. We must examine the question more closely.

We must try to generalize, but it isn't easy. It will be helpful to suggest certain group differences not so much between *microbes* as between *diseases*. Always remembering those blurred boundaries, we recognize classical infectious disease, thanks largely to Robert Koch, by certain sufficiently consistent features. Historically, the first of these is *communicability*. It is in the nature of infection that its agents— microbes, viruses, sometimes larger living things like worms—pass through the environment from one host to another. When they do so easily they produce outbreaks or epidemics. Contagion was known in ancient times, but the bewildering variety of its means needed analysis based on a definite germ theory and only began to yield up its secrets in the nineteenth century. The explicit germ theory of Koch goes on to tell us that a single species of microbe (or a virus) can be identified as the cause of each separate disease. The microbe is to be found in the diseased tissues, and—with exceptions—is to be absent in the healthy state. Grown outside the body and apart from any other living thing—that is, in "pure culture"—the microbe pro- duces in one or more animal species symptoms something like the disease in man. (This doesn't always happen, but it is very convincing when it does.) Antibodies that specifically match each microbe appear in the blood during the disease; demonstration of an increase in their concentration is of great diagnostic value. When specific immunity develops as well, it is a clincher; when it doesn't, the failure can often be accounted for by the existence of multiple "types" of the causative microbe, each of which gives immunity only against itself. And classically, in infectious disease, there is a characteristic interval of time between access of the microbe and appearance of symptoms— the "incubation period" of the disease.

These tidy logical points, so comforting to the orderly mind when they show themselves, are absent or all mixed up in the diseases we associate with the life on man, the so-called "endogenous" infections. There is hardly ever just one single species of microbe to deal with. Either several species seem to act independently in different cases, or a mixture of species, or more than one mixture, may do the job. The microbes found in the disease are the same as those found in health, a fact both obvious and confusing. The same, that is, as to species: but in disease there are either vastly more of them, or they are not in their usual place. Pure cultures of these microbes usually have little effect on white mice or other animals, or it may take enormous numbers of them to produce an effect. There are exceptions here, as I mean to tell you, and they are very important to our argument. But you can understand that the intrepid scientist is inclined to be contemptuous of these microbes: this kind of behavior is clear evidence of "harmlessness."

There are no significant antibodies; there is no rise in their concentration. The microbes, of course, have been around all the time. And there is no immunity; in fact these endogenous diseases have a nasty way of repeating, especially after we stop treatment with antibiotic drugs. And there is no incubation period, again because the microbes were there all the time.

Finally, and again logically enough, but confusing just the same, these diseases are not communicable. We already have the microbes without the disease, and nothing is contributed by getting more of them from somebody else. Thus the field of greatest triumph for public health, stopping the microbe in the environment on its way from one case to another, is closed to endogenous infections. So it is evidently not in any particularly practical way that we can speak of these normal microbes as "causes" of disease. But the difficulty is in fact verbal. Venereal diseases are not less infectious, nor are their microbes less causative, because the microbes are not in the environment long enough to let us scotch them there. Even when microbes seem to be the simple causes of infection, the appearance is deceptive. Causes of disease are never simple. They are in fact always multiple. Control is not always accessible by any given path; each problem must be solved in its own terms.

But although these normal microbes disobey the neat rules of Koch, there is still evidence that forces us to recognize them as disease-producers. Some of them can produce disease in animals in the laboratory. Endogenous infection responds to antimicrobial drugs. Germ-free animals do not have such diseases.

When normal microbes produce disease in laboratory animals, they do so under peculiar circumstances that are part of the pattern: we must not expect disease under normal or ordinary circumstances. An example is a disease of the heart—an inflammation of its inner lining membrane—that is deceptively called endocarditis *lenta* (slow), mainly to distinguish it from a more acute and rapidly fatal disease of the same place. The slower, or "subacute" disease is recognizably caused by many different microbes, often by a normally "harmless" streptococcus found in the mouth and throat. All but one of the negative conditions I have enumerated apply, and even that one seems to apply at first glance: pure cultures of the streptococcus in enormous numbers can be shot into animals without noticeable effect. The disease often occurs in people with damaged heart valves, damaged by earlier (healed) rheumatic fever or in some other way. When the heart valves of animals are first damaged, mechanically or surgically, or by exposing them to low atmospheric pressures ("simulated high altitudes"), subsequent injection of this otherwise harmless streptococcus causes the disease.

In man, given a bad heart valve, the disease has been known to start some time after a minor operation like a tooth extraction. It has been found that such operations can send a shower of microbes into the blood, which are recoverable from a vein in the arm within a few minutes. The healthy person mops them up quickly; it is as though such little accidents happen continually. But a roughened heart valve may catch a passing streptococcus, and it may get covered over with a blood clot before the white blood cells can reach it. Under these conditions it can grow; and the augmented roughness it generates attracts more clotted blood and the cycle continues. Bits of the "vegetation" break off and get carried by the bloodstream to capillaries or other tight places, where they settle, and the growth proceeds. The result is widespread disease, difficult to control even with miracle drugs, still often fatal today.

As far as we can tell, this streptococcus contributes nothing to which a morbid propensity for analogy could assign malice or any other human vice. It seems quite unable to maintain itself in the tissues of healthy animals. It is quickly engulfed and destroyed by white blood cells. It produces no poisons. But it can grow in clotted blood, and that seems to be all that is required of it to produce this fatal disease. Given the "opportunity" provided by presence around a tooth to be extracted (or a tonsil scheduled for the snare) in a person with a roughened heart valve, plus the off chance of surviving to settle on the valve and being covered with a blood clot—given this complex opportunity, the "harmless" streptococcus can start a rampaging disease that may finish off the patient.

This disease would not develop without the microbes, even though many other "causes" contribute to it. Yet the nature of the process is not fundamentally different from that in bubonic plague, in which rats, fleas, climate, and man's social peculiarities give the bacillus its "opportunity." And so it is with other endogenous infections, as we shall see.

Evidence bearing on the ability of normal microbes to cause disease that is likely to seem most convincing to the sick patient is the effect of antimicrobial drugs. Prompt relief of symptoms after taking medicine is probably the best-known example of apparent cause-and-effect, and, as most of us learn by experience, notoriously subject to fraud. Even the competent clinician's judgment can be faulty in such things. In recent years the "double-blind" clinical experiment, in which a drug and an innocuous "placebo" are alternated in test patients and coded, neither the one doling them out nor the one taking them knowing which is which, has been meticulously worked out to bypass errors of judgment. Not many such experiments have been tried with endogenous infections, and when they have, the results have been peculiarly mixed: unexplained good results with the placebo! Nevertheless there is a clear antimicrobial effect, and the trend of the evidence is unmistakably favorable rather than opposed to the idea of infection.

Another approach from which we might have expected more than we have received is the study of endogenous infection in germ-free animals. The difficulty here is in setting up experimental conditions *other than infection* to match those in man. Even if we knew exactly what all the conditions are for any given disease—and there is no instance in which we do—it would not be easy to contrive them; nor could we expect the conditions for animals to be the same as those for man. An experiment of this sort has been done for tooth decay in rats, and as far as it went, the result supported the idea that microbes are indispensable in this disease. Otherwise we must be content for the time being to know that, miserable as these germ-free animals usually are in their artificial purity, they don't develop anything resembling endogenous infections while their germ-free state remains intact.

All the information we have, then, tells us that the life on man helps us in the normal course but can hurt us in the abnormal one. Some examples and details remain to be examined.

Consider first *acne vulgaris*, common acne, the pimples of the face, sometimes of back and chest, so frequent among adolescents. In its near-universality acne shares honors with tooth decay. It begins as a *comedo* (plural, *comedones*), which when open to the surface is recognized as a blackhead. This is the plugged duct of a sebaceous or oily skin gland, which becomes inflamed in acne, forming the typical papule, or pustule. The microbes in the pus are the twigs and cocci of healthy skin, last seen by our own Gulliver. They are present in the comedo too; and in this early stage their numbers are greatly increased as compared with the normal. In the pustule the numbers are again reduced, pointing to the antimicrobial effects of the inflammatory cells. In healing pimples there may be no bacteria at all.

An experiment reported in 1963 by two Philadelphia dermatologists, in which cultures of the twiglike bacilli and the cocci of skin were injected separately into sterile sebaceous cysts in man, showed that the bacilli could produce inflammation and pus. It is almost co-incidental that this microbe has been called the "acne bacillus" for many years. Alone or with others it is probably responsible for the infection; but some experts still dispute or disparage the idea. Part of the difficulty, aside from the usual complexity of underlying causes, is the effect of antibiotics and similar drugs. Good results have followed their use, although, as we would expect, the good effect does not persist after the drug is discontinued. But in double-blind experiments such good results have also followed the use of the placebo. This and other parts of the acne pattern suggest an emotional factor in the illness. But the antibiotic effect is recognizably present over and above that of the placebo.

Here, as with other endogenous infections, the ruling practical question before us is, what can we do about this disease? Since the microbes are always with us, we can't hope to accomplish much by fighting them. At most we gain a temporary victory; and there is the lurking chance of doing harm to the patient by knocking out his normal microbes. Disease control is more likely to come from getting at the underlying "causes." But that is not likely to be easy.

The basic problem seems to be an overproduction of sebum, which on one hand provides nourishment for the skin bacteria (especially the twigs, which require fatty substances as food), and on the other may also furnish some of the protective effect of skin against infection, perhaps in the form of fatty acids produced from these substances by the normal microbes themselves! The attempt to restore balance by removing excess oil, with soap and water or with such fat solvents as alcohol or acetone, while often recommended in treatment, has recently been shown as well to allow *increased* bacterial growth on the skin.

There is much evidence that the sex hormones contribute to the development of acne, which becomes complete only when they reach adult concentrations in the blood after puberty. Hormone treatment has sometimes been effective where antibiotics were not. One suggestion is that the amount of glucose in the tissues may be important. And, as I suggested before, emotional factors can be neither ruled out nor precisely defined. Puberty is a time of emotional crisis in any event; and in the context acne is in danger of being thought of as "normal." But disease is never normal; we go on looking for more knowledge as the key to control.

I mentioned before that biological boundary lines are likely to be blurred. As an example, the boundary between infectious disease due to outside pathogens and endogenous infections is occupied by some that are hard to place squarely in either camp. A prominent group are those caused by the more pathogenic staphylococci. I refer now not to the cocci most numerous on healthy skin but to others, easily distinguished from the typical normal ones, and in fact showing all the marks of active pathogens except that they so often appear normally, on the skin, in the nose and throat, and in the alimentary tract. If we use the word "normal" in a statistical sense alone we must include these staphylococci among the forms of life on man. The diseases they produce are not typically endogenous, often starting, for instance, predictably after the microbe is transferred from one person to another. But I speak of them as borderline. They too seize "opportunity." They produce a pneumonia complicating influenza (a virus disease), and the complicating microbe is here more serious, more likely to kill, than the virus. And among other things, they cause a violent intestinal disease, an enterocolitis, in which the opportunity is provided by antibiotic drugs that destroy normal microbes which have otherwise, it seems, been holding the staphylococci in check.

But we are more concerned here with disease inside the boundary, caused by microbes with no particular reputation as pathogens. Such microbes, usually in mixtures of species, cause the inflammatory aspects of the commoner chronic respiratory diseases, including bronchitis and its serious sequels, bronchiectasis and emphysema. In bronchiectasis, parts of the bronchial tree have become distended and inelastic through damage to the underlying tissues. In emphysema, the terminal air-sacs of the lungs have been stretched and broken, and air has penetrated into the tissues themselves, destroying the breathing function of the area and so diminishing the capacity of the lungs as a whole. Among underlying causes of the whole array are cigarette-smoking, inhalation of other air pollutants, occupational exposure to irritants including coal and stone dust, and similar side-effects, especially of urbanization. These diseases have been described as a health problem today of potentially epidemic proportions. Endogenous infection is often continuous over many years, with repeated acute flare-ups, abetted by infections with outside microbes or viruses and by allergies. Antibiotics have value limited to individual episodes. It may be taken for granted that the direct chemical damage from smoke and other pollutants is considerable without the help of microbes, but it is the microbes that contribute the major symptoms of the inflammatory tissue-destruction and the pus-laden cough. Control is likely to be a matter of social engineering. The principal "cause" of these diseases is obviously man himself.

Some of the commonest and most troublesome endogenous infections appear in the intestinal tract. Special obstacles, among them the too-abundant mixed microbic population, have blocked the working out of the details here; and since the practical question of control, as before, leads one away from the microbes, the quest to incriminate special ones among them has been all but abandoned. Let me mention ileitis, appendicitis, chronic ulcerative colitis, and infection of the gall bladder, or cholecystitis. There is little doubt that the inciting microbes of all these and more are among the normal life on man. Acute appendicitis responds easily to surgery, and its complications to antibiotic drugs. For the others the continuing search for practical "causes" concentrates on such questions as allergy, hormone disturbances, and emotional upsets.

Peritonitis, following perforation of the gastrointestinal wall from any cause, which spills the microbial contents of the gut into the normally sterile abdominal space, is another example of endogenous infection, although extraneous pathogens have sometimes been accused of causing it. Peritonitis is usually preceded by one of the diseases of the gut itself. It is likely to be most serious when the perforation comes from a gangrenous spot in the gut or one of its appendages, like the gall bladder or the appendix. The gangrene contains an enormous overgrowth of some of the normal gut microbes, often particular mixtures which we know from experimental studies to be able to cause rapidly fatal disease when introduced into a sterile place like the peritoneal area. The once terrible mortality from this kind of accident has been all but abolished by the use of antibiotic drugs.

It is curious that one of the bacterial species often found as harmless in the intestinal tract is the poisonous agent of gas gangrene, notorious complicator of war wounds. One might think that any break in the intestinal wall, even an accidental nick during surgery, not to speak of the grosser cutting entailed in removing a section of the tract, would let this gas bacillus escape into the clean tissues with disastrous consequences. That this happens relatively rarely can be explained by the odd fact that this bacillus is a true saprophyte, growing only in dead or badly damaged and devitalized tissue. Like important members of the microbic mixtures in other kinds of gangrene, the gas bacillus has the primitive trait of growing only in the absence of oxygen. It has been reported that infectious complications of "potentially contaminated" abdominal operations have been greatly reduced by the simple expedient of delaying the closure of the wound, presumably by permitting free access of air to limit growth of such anaerobes.

There has been much interest during recent years in certain urinary-tract infections that are clearly endogenous. Several groups of observers put forward the hypothesis that large numbers of bacteria in fresh urine are a sign of trouble even though none is otherwise suspected. As we know, the urine is sterile in the bladder but picks up bacteria as it passes through the urethra. These workers reported that when they rejected the first part of the flow, letting it wash out the tube, and examined "midstream" samples, more than 10,000 to 100,000 bacteria per cubic centimeter was likely to mean disease somewhere in the urinary tract. The bacteria turn out to be those that live normally in the nether portion of the urethra. With occasional exceptions, no importance is attached to the *kinds* found, only to their *numbers*.

High bacterial counts in urine have appeared especially in women during the early months of pregnancy, in infants and young children, and in the aged. In the pregnant group especially, the high counts often persist through the months after childbirth and have proved to be associated with complications. The commonest of these, considered the source of the high bacterial counts, has been kidney disease, incipient and symptom-free in early pregnancy, emerging as obvious disease in the later months. What seems to happen in these women, as well as in the children and the older patients, is concerned with structural or functional abnormalities which interfere with complete emptying of the bladder and permit small amounts of urine to be drawn back into it from the contaminated urethra. Bacteria may be sucked back repeatedly this way, and establishment of the disease may entail additional outside infections—healing with residual damage —or other kinds of injury, since from experimental studies in animals it is known that it is not easy to infect the healthy urinary tract. The presence of such underlying damage is also suggested by frequent failure of these cases to respond to antibiotic treatment alone. Treatment being used in children includes surgical correction of the abnormality if possible, and the use of antibiotic drugs over long periods. These measures are recognizably imperfect, and no satisfactory solution of the problem has as yet presented itself.

Among many other conditions that might be mentioned in this sampling of endogenous infections are two that are very rare in our part of the world; they reflect the extreme poverty and hardship prevalent in other places. They are diseases of chronically malnourished children—diseases of starvation, to be blunt about it— called kwashiorkor and noma. Each is a horrible example of what the normal microbes can do, given extreme provocation. Kwashiorkor is found especially in tropical Africa. It is due mainly to extreme deficiency of protein, but multiple vitamin deficiencies can't help being present too. It begins with excessively prolonged breast feeding by an exhausted and undernourished mother—"breast starvation"— and continues under the regimen of a diet of local cereals without milk, green vegetables, or any source of animal protein or fat. Among the symptoms are some attributed to alterations in the composition of the microbic population of the gut. Details are not clear; but one would hardly expect the microbes to be in balance when the food *they* receive is so poor. Many of the deaths that result in spite of attempts at good dietary treatment—45 per cent of one series of hospital admissions at Cape Town—are evidently due to infection. Some of this mortality is found to be the effect of pathogens, but much of it appears to depend on the microbes that would be harmless in a well-nourished infant. Supplementing dietary treatment with antibiotic drugs has led to striking improvement in both symptoms and mortality. But obviously this is a disease to be prevented rather than cured.

There is no more tragic illustration of what "cure" may mean than that seen in the second of these diseases, noma or cancrum oris, a dreadful gangrenous ulceration usually around the mouth. It is seen mainly in children, an accompaniment of famine, malnutrition, and their infectious and other complications. It is one of the complications of kwashiorkor, and is also found as a companion of extreme poverty and wretchedness not only in Africa but in Asia, India, the Mediterranean countries, and Latin America. Many cases of noma were found after World War II among the inmates of the Belsen concentration camp in Nazi Germany. We know it to be the extreme consequence of spreading infection with the mixture of actively moving bacteria found in the gum clefts, which I have mentioned before and will speak of again a little later. These microbes are members of the normal population of the healthy mouth and throat and also of other parts of the alimentary tract. They proliferate in certain abnormal states, and when they do, they aggravate the abnormality and therefore tend to go on proliferating unless the whole process is stopped by suitable treatment. In noma the breakdown usually starts around a gum cleft; but with the tissues devitalized by the consequences of the wretchedness I have mentioned, it gets completely out of hand and spreads to take in anything in its path—mucous membrane, bone, connective tissue, muscle, and skin. And so it produces enormous gaping holes in the face, and keeps on going until, together with the damage done by its underlying causes, it leads inevitably to death—unless treated.

It was shown fairly early that treatment with penicillin stopped the infection, and when the underlying nutritional and other problems were also brought under control the patient could be "cured"; but, as one observer described it,

> the gangrenous destruction of tissue may be very great and leave a hideous deformity, for which plastic repair is quite impossible under tropical African conditions.

No adequate repair for the hideous deformity of poverty and wretchedness has as yet been found anywhere. Since we understand how these things come about and could easily prevent them if we had the means at our disposal, in this possibly most terrible onslaught of the normal microbes, ought we to blame them for it, or ourselves?

Let me round out this account of endogenous infections with one more example, pyorrhea. The word, which means "a flow of pus," is not descriptive. Pyorrhea is likely to become part of the experience of everyone who lives long enough, unless he loses his teeth earlier in life from decay or other things. But more than anything else it will have been pyorrhea that determined the "sans teeth" of Jaques. This would seem normal, part of the inevitable ultimate decline. But pyorrhea is disease, and disease is pathological, the opposite of normal in the intended sense. If death is equally so, and we can no more expect true prevention of one than the other, we can do our best to defer both.

The dentist usually tells his patient he has pyorrhea when his fine probes penetrate the gum clefts without eliciting pain or drawing blood. There are spaces—"pockets"—between gum and tooth root. Later the teeth become loose, and the upper incisors, for example, may be pushed forward and separated. There is a typical bad smell in the mouth. There is usually no pain, and the disease has developed so gradually that the patient may be unaware of it.

The course of pyorrhea is understood, although many details are still unclear. A complex and delicate balance of structures and forces, normally changing with time, is disturbed at any of several points; the microbes of the gum cleft, normally few in the free space, grow out of bounds and contribute to the disturbance; the process may be intermittent with episodes of arrest and healing, but the disease tends to progress.

As the teeth erupt in early life they are guided into place by the surrounding pressures of neighboring teeth, cheeks or lips and tongue, and the cusps of the teeth in the opposite jaw. As they come into function the propulsive force at the tooth root remains, and eruption is not stopped but only greatly retarded, so that as the grinding facets of the teeth are milled down ever so gradually, they are kept in functional contact. As part of this process the bone of the tooth socket is continuously maintained by the thin cushion of ligament that separates bone and root. Just as in the bones of the skeleton, bone growth responds to forces transmitted through such ligament fibers—here, from the impacts of chewing; in the long bones, from the pull of muscles. At the neck of the tooth a thin layer of surface epithelium dips over the bony rim of the tooth socket and joins the tooth, forming the gum cleft. The point of attachment must move as the tooth keeps erupting. This whole area—the thin edge of bone, with the outermost fibers of the root ligament, the connective tissue over it and the epithelium at the surface—seems to be one of special weakness, providing more than the usual set of opportunities for mischief by the microbic population, the active elements of which live on the root side, in the cleft proper. That is where the trouble starts. Here, in health, the toothbrush, toothpick, or dentist's probe elicits pain and bleeding if applied too vigorously. And as I said, any damage more persistent than such a pinprick may give the microbes their opportunity. An undue accumulation of food for them is provided by damaged tissue that cannot heal or be properly scavenged by roaming white blood cells, and by unremoved debris of the food we eat.

Normally the pinpricks always heal without trouble. Epithelial scales are sloughed off with dead microbes, and those not swept by muscular movements and currents of saliva back to the swallowing zone between tongue and soft palate are taken up by wandering cells. The configuration of teeth, gums, lips, cheeks, and tongue, lubricated with saliva, is beautifully designed to keep itself clean in the course of function. Food particles and dead cells do not accumulate unless there is a defect or disturbance somewhere. And the disturbance is most likely to happen at the gum cleft.

But though I speak of weakness—and man is, after all, "no more but . . . a poor, bare, forked animal"—let us not be too quick to disparage the design. This apparatus of teeth and bones and facial muscles with all its accessory parts is made for hard, long service. It is prepared to function several times a day, with dozens of poundings and hammerings each time, for twenty-five thousand days or more. Function is as necessary for it as it is for the muscles and bones in other parts of the body, or for the heart, or for every body cell. I have seen reason to believe that our teeth and jaws can in fact last a long lifetime in good repair: the teeth of Eskimos unspoiled by the white man's civilization, on the west bank of Alaska's Kuskokwim Bay, in 1936. They added to the usual use of their teeth the chewing of leather to soften it. They kept their teeth until they died. (But even so, some of the oldest ones were not free from pyorrhea.)

There is no single, specific, or invariable inciting "cause" of pyorrhea. Anything that interferes more than momentarily with the orderly succession of changes at the gum cleft may do it. We see such interference happening, more quickly than in pyorrhea and with different results, tending to heal and disappear but probably always leaving residual defects, in many circumstances. In nutritional upsets, especially in scurvy; in certain cases of poisoning; in such childhood virus diseases as measles and a first attack with herpes ("fever blister") virus—in all these the mixed microbes of the gum cleft are given a chance for a rampage, subsiding as the underlying disturbance is corrected. The rampage is accompanied by inflammation directly due to the microbes. This happens as well when the surface tissues are damaged slightly but persistently by fillings or bridgework poorly made, with tiny gaps or projecting edges. As we reconcile ourselves to accumulating evidence we see that emotional crises can be accompanied by mouth disturbances of this very sort; and we now attribute acute flare-ups of gum disease in soldiers and students to this kind of origin, with accompanying bouts of fatigue, alcoholism, and hot-dog-and-Coke diets.

But such things give rise to acute, clinically obvious disease of the gums that is not pyorrhea. The residual damage they may leave on healing, and less obvious effects induced by more trivial events, are surely the starting points for the set of changes that emerges later as pyorrhea. Somewhere in the course of this process there is an accumulation of food for the microbes in a protected zone that remains unrecognized and unremoved. Slow, insidious damage affects the bone at the edge of the socket and the epithelium at the point of attachment to the root. The relationships are disturbed and a pocket is formed. Sloughed scales, tartar, and residual food find their way into the pocket, and the gum-cleft microbes are given the basis for riotous living. The result is the profusion of spirochetes and other microbes I have spoken of before.

We know from experiments in animals that the mixture of microbes found in these pockets, when scooped up and injected under the skin or into the peritoneal space of healthy animals, can produce some of the symptoms of pyorrhea—or of noma: the difference between these diseases is a matter of degree. Such animals show the same sort of inflammation and tissue damage, the pus, the characteristic stink, and the rampaging spirochetes and other microbes. We need not doubt that the microbes contribute most of the damage in pyorrhea, even though it is plain that they can do nothing without the other "causes" of the disease. Treatment must concern itself with these other causes; antimicrobial drugs are either very temporary in effect or useless. Careful cleaning of the pocket with delicate instruments and rigorous but competent toothbrushing can slow the process down. Surgical removal of gum tissue to eliminate the pockets may arrest the disease. These measures may keep the teeth in function and re-store the sweetness of the breath provided that neglect has not let things go too far. "Cure" follows removal of all the teeth, spontaneously or via the dentist's forceps. The method differs only in magnitude from curing a headache with the guillotine.

A final word is called for on something else our microbes do, not so much for us as with us—a piece of work of beneficence at long range, although we might be deeply offended by it if we were not beyond such sentiment. Once we are dead—and indeed before we actually die when we do so little by little, in terminal illness rather than sudden accident—the barriers that kept the microbes outside in life come down, and the work is begun of returning our substance to the economy of nature. What happens thereafter is known less from direct studies of necrobacteriologists—at all events, I know of few such—than by piecing together much test-tube knowledge. There must be great changes among the normal microbes, with emergence of those suited to the lowering temperatures and a now frankly saprophytic situation. It is hard to say at what point the ministrations of these descendants of the life on man are augmented by those of the microbes of soil. No matter; we may rest assured that our microbes, in the best interests of posterity, perform these last rites for us.

14

Obscenity Reconsidered

> . . . where we generate disease to strike our children down
> and entail itself on unborn generations, there also we breed by
> the same certain process, infancy that knows no innocence,
> youth without modesty or shame, maturity that is mature in
> nothing but in suffering and guilt, blasted old age that is a
> scandal on the form we bear. Unnatural humanity!
>
> —Charles Dickens, *Dombey and Son*

Nothing is so difficult, I think, as to speak plainly—to have something to say, to say it simply, and to get in return the validation of being understood, simply again, but completely as well. It is one of the beauties of science that it can do this within its own limits. Scientists can speak to their fellows anywhere in the world, and if they stick to science they can expect to be understood, and if they stick to verified data they can expect to be agreed with. Religion, politics, skin color, ethnic background have nothing to do with it. Science is basically without prejudice. Stripped of prejudice, it gets the message through. American viruses, Chinese insulin, Indian equations, Russian Sputniks, Japanese electron micrographs—the nationality is irrelevant. Much weaker links join the graphic arts and music; still more tenuous ones stretch between nations in literature and philosophy. Great yawning chasms separate peoples—continents, nations, even different communities in a single city—on every other subject I can think of.

But science by itself is not of much use, as we have been learning. Even complete agreement among scientists of the world could not bring peace on earth. The scientists would agree, of course, only in matters of science, in fact only within the narrower limits of their specialties. Outside of science, as we have discovered since 1945 or so, they disagree as much as everyone else. Away from their special areas of competence, they have prejudices just as non-scientists do. Occasionally they even let prejudice color their opinions within science; I have suggested a few examples.

I have been talking about the interplay between a particular piece of science and a particular set of prejudices. It has been my idea that if the two were brought together the science might shine a light on the prejudices and show how foolish they are. I have taken it for granted that I had something to work on, that we already suspected they were foolish, that my job was not so much revelation as encouragement.

We are in a mood today to re-examine old notions and prejudices. Young people especially are challenging the accepted ideas, and they have already decided that some are foolish and have abandoned them or are in process of abandoning them—both prejudices and consequences. Hence there is a revolution in progress—sexual, sartorial, verbal (and in China, "cultural"!), and certainly more. Some of it is peaceful and may even seem reasonable to non-revolutionaries; some of it is alarming. My sympathies are with the young; but my objective is rationality.

The prejudices I have been attacking radiate out of the core idea of obscenity, which we talked about along the way, in Chapter 8. It is time to return to it. I suggested there that the problem is one of disease. This is what Dickens intends to point out in the quotation that heads this chapter; his key words are "modesty" and "shame." I would like to be both radical and conservative: I want to lift the veil but not to tear it to shreds. Nothing need be hidden or denied, I suggest; there is good, and no harm, in embellishment and emphasis. We ought not to lose delicacy and good taste. Even words ought not to be used to draw blood. Ought we to keep the modesty and throw away the shame? Modesty—the genuine kind—is humility, simplicity, unostentation. Shame turns humility into humiliation and leads on to guilt and disgrace. Perhaps there is a place for such feelings, but they ought to be used with more discretion—for adults only?

We are coming to agree generally that nothing in healthy sex needs the protection of shame or of anything covered by the idea of obscenity. Healthy sex is not shameful, not obscene. Granted that we don't always know where health ends and disease begins—this kind of health like other kinds. We are to learn more as theory and practice both grow with experience and as the biology of sex is taught more and more openly to children. I think this part of the problem may work itself out. Not that it has done so yet by a long shot. Our ideas, for instance, of what we need to protect the very young against seem to me still rooted in mythical fears.

I want to dispose of the matter of sex but not to dismiss it lightly or flippantly, so let me say a few more words about it. I think the core of the sex problem today lies more in actions than in words, and less in the young than in what their parents do. Where there is reckless promiscuity among the young, from which sprouts the ugliest branch of the problem—venereal disease—the wild growth explodes under restraints misapplied by cynical and mendacious elders rather than from ill-advised freedom. Trace the cynicism and mendacity back to their roots, and I think you will find the mythical fears. Most experts now agree that the VD problem will crack in the full light of knowledge; and now that unwanted pregnancy can be virtually abolished, we could probably relax and watch the whole problem evaporate if we could come to agree that this is all there is to it. Youth is not free from its own distortions, apart from venereal disease. But there is enough responsible affirmation among young people, I think, to make them turn out better than we did if we let them alone.

In 1968 the Supreme Court for the first time approved a New York obscenity law that applies only to children, prohibiting the sale to persons under seventeen, as reported by *The New York Times*, of " 'girlie' magazines and other literature depicting nudity, 'sexual excitement, sexual conduct, and sado-masochistic abuse.' " Justice Brennan, in handing down the 6-3 decision, the news report continues, "noted that most experts doubt that nude pictures and girlie magazines are harmful to children. But he said it was 'not irrational' for the Legislature to find otherwise." Presumably the double negative implies that it isn't rational, either, or he would have said so. This looks to me like censorship in retreat, fighting a rear-guard action. It is evidence, if evidence is needed, that the revolution is still in progress.

The old notion of obscenity is one of the things we must get rid of. We might aim to do away with the word entirely, but, at best, that will take time. One can't just extirpate a word from the language and burn it, like a book, or see it shredded in the garbage disposer.

Even the idea of such an operation is repugnant. Can't we conserve this word and find a new use for it? And if doing so makes me a conservative, I see no harm in the label.

Words have disappeared from language, but I doubt that any word has ever been abolished by choice or plan, or by what amounts to the use of force. So our choice with the word *obscenity* is either to try to stop using it, perhaps to question it whenever we can as it is used by others—and to hope, quite likely vainly, that this will do some good—or to use the word in a new way. Dictionaries follow usage, so this second course may work. I have already suggested what the new meaning ought to be, and of course others have done so too: I would assign as its literal meaning one that is now often applied figuratively—not just sex or scatology, *maybe not sex or scatology at all*, but any offense to decency, to human values we call moral, values we need to preserve us. This is large territory, and I don't intend to explore more than my proper corner of it. Maybe the rest can be left to those theologians who are willing to renounce the last remnant of Puritanism.

If there is nothing ugly or reprehensible about man except what can be defined as disease and treated accordingly, we ought to separate the notion of obscenity from all healthy human acts and functions, and from the parts of our bodies concerned with them. Words descriptive of those parts and acts will be the first things to be liberated. Restrictions based on the origin or length of the word will have to go. It must be permissible to speak of procreation or excretion with as free a choice of words as we can use in speaking of execution, war, or the bodily function of a jet airplane. (In a TV commercial I have seen a jet plane farting across the screen; is this supposed to be beautiful?) But forgive me if I continue, with only occasional exceptions, to favor the longer aristocratic words; I am beckoning toward Utopia, but we haven't arrived there yet. Besides, it is easier to say than to do; as much as anyone, I am bound by the habits of a lifetime.

The heretofore forbidden words, taken into the living room after having lurked so long in the latrine, can be expected to respond to the benefits of fresh air and exercise. With practice we will learn to say them with no more emphasis than we give words already tamed; and, as we do so, they will stop grating on our ears. They will come to take their place in our vocabulary as nouns, verbs, participles, maybe even as colorful metaphors—as we have retained the old usage of "spit" as a narrow projection of land running into the sea. But, losing the force of rebelliousness, they will stop being used as expletives and oaths, as verbal weapons. We will have taken the poison out of them, detoxified them. Waste no regrets over this; we are unlikely to be left lacking in weapons, verbal or other. Artistic invective has never depended on these words, and inartistic verbal cruelty—which will continue to come under the new definition of "obscenity"—will find other resources all too easily.

If this suggestion of mine can be accepted with all its implications, the detoxification will be complete and there will be no more harm left in the procreative-excretory words. The same ought to go for pictures as well, for anything having to do with healthy body parts and functions. The new notion of obscenity will take over where health stops and disease begins. But this change can come about only gradually; and before it is complete there will still be ways to use words or pictures conveying ideas of sex or scatology either deliberately to injure or for other selfish, distorted purposes, perhaps even merely carelessly, so that injury results or becomes likely. The distinction will never be sharp, as it never can be between health and disease generally. It will become easier, I am sure, as we shift the emphasis away from body parts and functions to things and acts that are really evil, and as we recognize that our immediate concern is with disease rather than with crime, calling for prevention and treatment rather than injunctions and tear gas.

Words may be used to build, to heal, to beautify, as well as to injure, to debase, to destroy. For either use it is not nearly so much the words themselves as the way they are used that determines the result. Take the quotation from Dickens at the head of this chapter as an example. This little passage, and especially the whole paragraph from which I took it, suggests what an artist can do with a limited palette. Goya in his etchings, and Daumier in his somber paintings, could also give us, as Dickens has done, a sense as though in full color of the "moral pestilence" rising out of a city slum. This whole passage tends to be scatological without abandoning any of its Victorian propriety, in fact even without conceding anything to Victorian hypocrisy, except perhaps for the one word *shame*. Whether Dickens could have done better with full verbal color is a question I cannot answer. Perhaps he might have been as great as Chaucer or Shakespeare if he had had their freedom of words. My enjoyment of Goya and Daumier takes nothing away from my love of Monet.

Come back with me on another visit with Freud. It was he, after all, who started the sexual revolution. Not the least part of the value I find in him comes from his emphasis on health as well as on disease. As a physician whose starting point was mental illness, whose best-known contribution was a form of treatment, he nevertheless continued to think of himself as a psychologist; and his greatest contribution may turn out to be the pattern of a healthy mind. He insisted that mental disease springs mainly out of distorted notions of sex and excrement, and in so doing gave us the basis of a healthy attitude toward both. He did not escape some of the prejudices of his time. He made class distinctions that seem invidious to us now, and he took for granted notions of "coarseness," "vulgarity," and "smut." But the meaning he intends comes through such things without distortion.

The part of Freud I have left for this chapter is his discussion of humor, especially so-called obscene humor. Freud uses the adjective in the old sense. It becomes clear as we read him that, like obscenity itself, the humor associated with it—especially the dirty joke—extends into the domain of disease. All I have said about words and pictures applies here, but Freud throws a spotlight on the whole question.

The special feature of the dirty joke, as Freud pointed out, is the explosiveness of the response to it. The sexual or excremental joke, he says, differs only in its subject matter from others, falling otherwise into similar categories. The techniques of telling it are also the same. Dirty jokes always have a motive, or, as he puts it, a "tendency," beyond that of being merely amusing or comical. They may be hostile, by which he includes the intent to be aggressive or satirical, or only defensive. Obscenity Freud equates especially with exhibitionism. "Smut" he defines as "excremental in the most comprehensive sense," that of the cloacal confusion of childhood fantasy. Any joke provokes laughter by liberating pleasure through a broken barrier of inhibition or repression. The greatest charge of pleasure, hence the most explosive burst of laughter, is released by breaking open the most deeply repressed material, which is excremental in Freud's inclusive sense.

The technique of the joke, and in part the material that makes it up, varies with the intellectual level of joker and audience. Freud says it is easier to provoke laughter with "smut" in vulgar people than in the educated. I think he implies that repressed material lies deeper, is harder to release, among the educated. As the joke technique becomes more refined or, as we would say today, sophisticated, the element of surprise, which Freud describes as leading the listener along a path toward something portentous or ominous and suddenly bringing him up short without pain or penalty, is more polished or subtle. But the basis of the dirty joke is repression, which is abnormal. We might still laugh at such jokes if we had no inhibitions, but only as we do at non-excremental jokes, not explosively. We would seem to have given up something here, as in the matter of swearing. But I wonder, first, whether we would not have gained more than we had lost, having exchanged health for disease; and whether, metaphorically, it isn't better to be able to keep the pressure from building up dangerously than to let it do so and have safety valves to keep from exploding. There are questions here, perhaps, that are not so easily disposed of. What happens to the really healthy-minded fellow when he bangs his thumb with a hammer? Maybe we will have to wait and see (or hear).

As an amateur in this field, I have given a little attention to what has been said about jokes, dirty or otherwise, since Freud's time; but I found nothing that seemed especially useful. A current book, *Four-Letter Word Games*, written by a psychiatrist for a general audience and purporting to cover the work of others on its theme, seemed to me bad. I mention it for the sake of serendipity. Its subject-matter is both words and people, and it struck me as being oddly lacking in sympathy for both. The people all seem to be presented as pathological, including both those described as patients and others; I got the impression of a one-eyed man in the country of the blind. One expects compassion and fails to find it. And the author consistently makes the use of "four-letter words" by his subject people abnormal; yet he himself strews his text with the same words, it seemed to me wildly, much in excess of any valid need.

The serendipity comes in the unexpected example completing the circle of words and meanings. Excremental words are not obscene in themselves, no more than any others; *but they can be as obscene as others* when, like the others, they are used in a mean or ugly way, carelessly, derisively, sadistically.

There is nothing new about all this, and I may seem to have come to it by a longer and more devious road than have others through the ages. It has been said, and is still being said, by youngsters whom we dismiss as innocent, whose truth is child's truth, unassailable and refreshing but unsuited to a cynical world. It has been said by Rabelais and Mark Twain, whom we dismiss as not really having intended to be anything but funny. It has been said by Joyce, whom we forced to speak a private language; and even so we tried to stop him. It has been said by Lenny Bruce, who was not a child, who spoke so plainly that when he made us laugh it was as we laugh at alienation, at Albee, at Pinter, who are not at all funny. He was a witch, like Joan of Arc; and we killed them both. But what all these people said needs to be said again and again until the message is finally delivered.

We must free our language, free all of it, so that we may learn to use it, using it with respect and love, as we must learn to do with all freedom. The less skillful we are with words, the more words we need. This freedom is more important for ordinary people than for artists. Dickens could convey his meaning in a verbal monochrome better than most of us can do with a full palette of words. Alec Waugh, the novelist, looking back through fifty years, in *The New York Times Book Review*, argues that the new freedom to use four-letter words came late for him. He "had learned to convey the most devious undermeanings surreptitiously" without them. As Greek drama left violent action offstage, he suggests, we ought to leave things to the reader's imagination. But it is one thing to practice exercises for the left hand alone on the piano, or the G-string on the violin, limiting the means to develop skill; ought we to forbid the use of the right hand or break the other three strings? Shakespeare probably used the least limited vocabulary in the history of English; yet nobody knew better than he did how to play on imagination with every device of indirection and subtlety, and nobody knew better how to speak to the groundlings as well as to the literate elect.

The revolution is afoot, going in the right direction, but stumbling. Some words have become permissible, others still are not; there are things the movies still can't say or do that the theater can—especially the further it gets off Broadway. People can discuss their sex lives on a television program: I haven't seen or heard it but only read about it; no doubt certain verbal restrictions apply, and the report I saw mentions that one must not get scatological or talk about bodily functions. In the movie version of *Ulysses*, which seems to have got by because it is almost as forbiddingly intellectual as the original, the spoken "raw, ripe Elizabethan words" are described by an eminent critic as not corrupting because they are given "superb articulation." In Sweden it appears that restrictions against pornography have been almost completely lifted. "Sex is a natural thing" there, and even the sex organs need not always be hidden, although they cannot be shown copulating; frank homosexuality is also frowned upon. Whether other bodily functions can be displayed, or whether, with so much freedom, nobody feels a need to display them, is not mentioned. In France, on the other hand, a play by Picasso (written in 1941 and staged in 1967 in a tent outside St. Tropez, where the mayor had banned it) presents a bare-breasted actress—"a professional stripteaser"—squatting at the center of the stage while a sound track gives forth a gushing, gurgling noise. At the theater in Minneapolis named for Tyrone Guthrie, the first act of a play attempts to set a tone of civic corruption by diffusing into the audience the actual ammoniacal reek of a public toilet. A revival of Brecht's *A Man's a Man* has a two-man mock elephant urinating in full view with the aid of a bottle of water emptied under it; and the hero, his legs showing under the partition of a public urinal, produces a dripping blot against its translucent surface. The Russian poet Voznesensky, repudiating charges in a newspaper story that he is anti-Soviet, is quoted as saying that "such allegations [belong] in a Paris *pissoir*." A boy doll with recognizable genitals makes its appearance amid much argument, a good deal of it approving. Nudity in movies becomes more and more common, promoted especially by avant-garde foreign films, which, however, also lead a trend toward sadism. A student literary magazine in Little Rock is censored by the Arkansas State auditor(!) because of poems that "referred to excretory functions." And LeRoi Jones is convicted in Morristown, New Jersey, on charges of possessing weapons, of which he is widely believed to be innocent, after the judge noted that he had used a word "descriptive of excrement."

Some of this is good and some is not, and it isn't easy to know for certain, precisely or in all instances, what is healthy and what is sick. I am in favor of the freedoms, even when they seem to be abused, and against the chains and padlocks. Give us light, and we will find our way. Censorship is retreating, fighting rear-guard actions. Lenny Bruce has been exonerated *post mortem*; maybe we will sanctify him as we did Joan. Apparently his efforts had "social importance" after all. A lawyer whom he dismissed before the end of the original trial now argues, as Lenny knew, that restrictions on English words began with the Battle of Hastings, that "the exact synonyms" of words he used, coming into England via France, are not considered offensive, including the words "fornication" and "excretion." A young actor, Peter Fonda, son of Henry, interviewed after making the movie *The Trip*, is quoted as saying, "There is no such thing as a dirty picture and there are no dirty words." He discounts the idea that children are damaged by seeing or hearing such things and contrasts them with lies in public affairs and advertising. The story is told indulgently, much as we might recount the bright sayings of youngsters; there is a hint that Peter is eccentric. Turn the page and read about the crisis of credibility.

Here is what five-year-olds are saying in England, according to a story reprinted from the *Guardian*. They are denouncing one another, evidently in a nursery school. Little girls seem to do the accusing. "He swore . . . he said 'underpants' . . . 'kidneys' . . . 'adenoids.'" A distorted mirror held up to parental injunctions: the straight, clear logic of children.

Is it the truth that hurts children, or do we rather damage them by telling them lies, by hiding the truth from their eyes and ears? I would like to see the experiment tried of bringing up a group of children without any hypocrisy at all, including the euphemisms by which we try to make hypocrisy itself look like something else. To do this right we would need not only the biology of sex, starting as early as possible in ways the experts are now working on, with science and art together, truth and beauty. We would need as well the microbiology of the body parts and functions, again the art as well as the science. I have been groping toward something of this sort in the devious course of this book. One has to begin with adults, especially with parents. They will need to translate it for children, modify as required, add and subtract. I offer you a prospectus, not a blueprint. The job will be difficult; but the time is ripe for it, and it ought to be tried. It ought to be done modestly, with all possible delicacy and good taste: gently, but without any lying, secrecy, subterfuge; certainly without shame. We need not venture to predict the outcome before the experiment is tried; but we can be reasonably sure that, at worst, we won't be doing any more damage with the truth than we are now doing with our lies.

Consider one more example of the fruits of prejudice, bringing us back to microbiology. As science does not prejudge but lets the truth lead it wherever it may point, scientists ought not to be subject to prejudice. But they are people like others, brought up just as others are, lied to in the same ways. They are likely to have the same prejudices against the implications of the life on man that others have —associated, as we have already seen, with prejudices against the microbes themselves. One such prejudice, which I had formerly shared, or at least had not seriously questioned came to my attention while I was writing *Microorganisms Indigenous to Man*.

Part of the objective I set for myself in that book was to decide which among the dozens of species of microbes found on man really belonged there, lived there more or less permanently, were, in a word, indigenous to man, and which were only transients, including disease-producing foreigners as well as harmless ones. Among the microbes are a small group of species of protozoa, or one-celled animals, found especially in the intestine, but also in the mouth and the vagina. As you probably know, protozoa tend to be larger than bacteria and have more obvious complexities. Some of them move around, as animals tend to do, a few of them with a good deal of vigor, giving rise to much commotion in their liquid vicinity—like a great lashing, tumbling fish scattering all the little ones. This picture itself, the main association of these microbes with feces, the fact that, like bacteria, they have close relatives that produce serious disease: these are some of the elements out of which prejudice grows.

Most of the scientists who have studied these microbes are parasitologists, by which name we mean specialists in infectious diseases due to protozoa and larger animals, especially worms and insects. If there is anything that distinguishes these people from specialists in other areas of biology, it is likely to have something to do with their need to work with feces more than other biologists do, in fact more, or more intimately, than any other group of people I know. Feces is the natural habitat of many of the species they work with, and a good deal of their time is spent preparing specimens of feces and looking at them through a microscope. This is their work, and they quickly overcome any special feelings you might think they would have and go about it in a matter-of-fact way. But it gives them something of a scatological bias, something like the good-natured eroticism of venereal-disease experts. Both may tend toward defensive ribaldry at times, for example when they and their business are first introduced to strangers. In appropriate company, especially with a new class of medical students, the parasitologist is likely to use what may possibly be one of the oldest jokes in English. Reynolds mentions it, and it wouldn't surprise me to learn that it could be traced back to the gong-fermors of the fourteenth century: "It may be shit to you, but it's my bread and butter!"

These scientists have taught us most of what we know about the protozoa that live on man. But, like bacteriologists, the parasitologist is much more interested in the obviously pathogenic protozoa than he is in the group whose credentials I was looking into to see if they could pass as citizens. These pathogenic protozoa, like the intestinal worms, tend to be transmitted from man to man in feces, and the pattern of their presence or absence is related to habits of personal and group hygiene, hence to various associated attributes of what we like to call civilization or the lack of it. It is not hard to see how the combination of feces, filthy living habits, social backwardness, and all the rest might have nourished feelings of prejudice, but parasitologists have resisted any such tendency and have on the whole concentrated objectively and effectively on the great problems of public health. They are probably the least bigoted among the advanced people who have had contact with backward people throughout the world. But if the prejudice-generating combination with which they work was deflected off the people they work with, it fell heavily on the microbes, especially the ones the parasitologists were likely to be contemptuous of, the relatively unimportant ones in the usual sense of involvement in serious infectious disease. And so it came to be lightly assumed, and the idea came to be passed down from one textbook writer to another—accepted without re-examination because other matters were more pressing—that the non-pathogenic protozoa, like the disease-producers, are associated with filthy living conditions and social backwardness. I was taught this as a student and didn't think to question it until I came to it in writing my previous book. If it is true, if protozoa are found only or mainly in filthy people, they are not indigenous. Is it true?

I spent a good deal of time exploring the literature for an answer to this question and didn't find one that is entirely convincing. But this would seem to be like a question of guilt, and insufficient evidence ought to lead to acquittal. In more scientific terms, we begin with a null hypothesis and do not accept a relationship—between filth and non-pathogenic protozoa—unless we find compelling evidence for one. I spread the evidence over fourteen pages of *Microorganisms*, compiled it in seven tables, and found it short of compelling any conclusion. The long-standing contention of the parasitologists is certainly not borne out by the facts. There is suggestive evidence of a false conviction by association—of non-pathogens with pathogens—and, in an oblique way, an association with *disease* rather than with filth or backwardness, it being noted that some connection exists among all these things.

The acquittal is justified for the intestinal protozoa themselves, but the evidence having to do with the slightly different protozoa of mouth and vagina, to which the textbooks apply the same prejudice, points more directly toward disease as the culprit. One cannot, after all, escape the fact that feces is spread around with less restraint, or more abandon, among socially backward than among technically advanced peoples. Protozoa tend to be found more often where there is no plumbing. But not consistently: backward peoples in tropical and subtropical regions show them prominently; but natives of Alaska and Greenland, no less innocent of sanitary conveniences, show a pattern of intestinal protozoa more like our own. Attempts to associate hygienic habits directly, separating them from associated circumstances, with intestinal protozoa (others before me having questioned the conventional notion) had all led to ambiguous findings. A single study, made by Asa Chandler in villages near Egypt, was made of comparable groups of people before and after the introduction of Western-type latrines and pump wells. There was a clear diminution of intestinal *worms* in the group with plumbing, but no shadow of a difference in the prevalence of non-pathogenic protozoa. But the horse-to-water adage was introduced: there was, in fact, some evidence that the natives preferred not to foul the shiny new equipment and continued in their old ways. And the two-year period of their existence may not have been enough. It is nevertheless true that many people who have consistently observed the niceties provided by modern sanitation nevertheless have these protozoa in their guts. I have accepted them as indigenous, subject to loss of naturalization on proof of guilt.

With the protozoa of mouth and vagina we come closer to home, since we need not stray outside one country—and the United States serves as well as any—to find all the evidence we need. Here a stated association with cleanliness or its opposite is quickly seen to be figurative rather than literal to begin with. A "clean" mouth is in fact a healthy mouth; and in the genital tract "clean" becomes the reverse of an illness thought of by many as one of the minor venereal diseases; but transmission by sexual contact is denied by others, who think of the microbes as indigenous and the symptoms as comparable to those of diseases I spoke of in the last chapter. In the mouth, in fact, related protozoa—called trichomonads—are found mainly in association with pyorrhea, to which, nearly all students of the subject now agree, they make no significant contribution. These microbes live on bacteria and seem to flourish in pyorrhea pockets because of the rich supply of bacterial food they find there. In the vagina there seem to be underlying causes too, but they are less well understood; and the vaginal trichomonads have disease-producing proclivities lacking in those of the mouth.

Both pyorrhea and trichomonal vaginitis are more common among the lower social strata of American society. The vaginitis has been associated more often with clinic patients than with private ones, and more with blacks than with whites. You will understand that these circumstances are sometimes suggested as arguments for the dirty-versus-clean hypothesis. But since the factor of disease is certainly present, and since the diseases themselves are not associated with cleanliness or the lack of it, at least in any way that is seriously thought of as incriminating, we may bring in, this time, a clear verdict of not guilty as far as filth is concerned. In the matter of pyorrhea it is not difficult to find the connection between disease and social condition in such things as nutrition, lack of effective early dental care, and similar features of poverty. The story of trichomonal vaginitis suffers from lack of information; but what we know is consistent with a similar pattern. I consider both mouth and vaginal protozoa to be indigenous; the disease pattern of the vaginal ones is typical.

Not filth, then, but disease. The distinction is fundamental. If you are filthy, you are yourself to blame; go and wash, or be ostracized. But if you are sick, we owe you compassion; you need a doctor; it is our moral duty to help you get well again. Brushing the teeth has some value in pyorrhea, but that is a special case. It is generally true that you cannot wash or scrub away disease. Disease is never shameful to the person who has it, although much of it ought to shame the society that fails to deal with it. Yet it can often be prevented by things the individual can do for himself. Keeping clean in appropriate ways is one of those things. But there is a limit to what cleanliness can accomplish; and when we overstep it—when we get overzealous about cleanliness—we can do ourselves more harm than good. We need to live with the life on man, and we ought to let well enough alone, which is the subject of the next chapter.

15

The Shell Game

> The results of our work . . . show that the organisms
> deposited on the rim of the communion cup are not destroyed
> within the short time—5 sec. as an average—elapsing
> between the partaking of the sacrament by each successive
> communicant.
>
> It must therefore be admitted that the common communion
> cup may serve as a means of transmitting infection. Reasons
> are given, however, for believing that the risk of transmission
> is every small, and probably much smaller than that of
> contracting infection by other methods in any gathering of
> people.
>
> —Hobbs, Knowlden, and White (1967)

Not many years ago every medicine chest could be expected to contain a bottle of tincture of iodine, and painting it on every cut or skin abrasion was a ritual. At first it was strong tincture (7 per cent); after a time it was reduced to 2 per cent. By now you ought to have thrown it away. Its use is no longer recommended. The accepted treatment for minor skin wounds is to wash them clean and keep them dry and covered with gauze that lets air in and keeps dirt out. Mercurochrome was used for a while and abandoned as, at best, useless. That is about all that can be said for antiseptics in general: they do no good, and they may do harm.

The tissues of a clean wound need no assistance in the healing process. Even if a minor wound gets dirty and infected, inflammation throws off the invading microbe with the pus that forms, and healing follows. You may be able to hasten the process by soaking the part in moderately hot water. If the infecting germ happens to be the kind of streptococcus that causes sore throats, the neglected little wound can become very serious, spreading and, if unchecked, sometimes leading to fatal blood-poisoning. Penicillin has made such tragic accidents rare. If the wound is deep, however small, like one made with a dirty needle, it may lead to tetanus, a calamity unless caught and treated immediately after the wound is made, preferably, in the vaccinated, with a booster shot. If it is a large as well as a dirty wound, the risk of tetanus is compounded with others, including gas gangrene. If the wound is part of hospital surgery, it sometimes gets infected with one of the staphylococci that lurk in hospital personnel. Less often, and especially when the hospital patient is a burn case, the wound may get infected with an environmental bacillus that causes blue pus ("pyocyaneus"). These are all currently serious problems: tetanus, gas gangrene, staphylococcus and pyocyaneus infections. They are serious partly because of difficulties in using antimicrobial drugs, in tetanus and gas gangrene because the main culprits are bacterial toxins the drugs don't touch, in the other two because the microbes tend to be drug-resistant.

Antiseptics have no significant part to play in dealing with these problems, with a single exception. It is now recognized that even the antiseptic used on skin in preparation for a surgical incision is likely to be ineffectual and relatively unimportant. The microbes that cause infections in hospitals (and elsewhere, for that matter) are not the ones found on healthy skin, even though disease-producing staphylococci are often found there. Exceptional instances aside, the offending microbes come from other people, including the operating-room personnel, or from the environment, including some of the equipment in the operating room that is hard to sterilize. Surgeons have been trying unsuccessfully to sterilize *their* skin and to get the staphylococci out of their noses and throats. They keep everything covered, instead, with sterilized cloth and rubber gloves, and try to sterilize everything that comes near the patient.

Antiseptics have been largely given up not only because they have been found generally useless—I am coming to the exception—but because their intended effects are accomplished much more efficiently with antimicrobial drugs. These are a special class of chemicals—or two classes. One group is made in the laboratory, like Ehrlich's arsenicals (starting with his "606") once used especially for syphilis and going back to 1910, and the sulfonamides, starting with Domagk's sulfanilamide in 1935, which initiated the modern therapeutic revolution. The other group, beginning with penicillin (discovered by

Fleming in 1928, but first made generally available by Florey and Chain in 1942), originally consisted of products made by microbes—hence "antibiotics"—but some of them have since been synthesized. All these antimicrobial drugs have a special selective action against a limited range of microbes and tend to spare the patient's tissues in which the microbe is growing. This rule is far from absolute, but it is true enough in practice to distinguish these antimicrobial drugs sharply from the antiseptics, which damage or destroy living matter indiscriminately.

The antiseptics range from such elements as chlorine and iodine, to simple salts like mercuric chloride, to increasingly complex organic substances like phenol and a range of its derivatives—and many others. They have no selective effect on microbes. It is a general rule with variations in degree that if they kill germs they also damage or destroy tissue cells. For a while we thought we had a group of exceptions among certain detergent-type agents, and great hopes were pinned on them when they were introduced; but they have since been relegated with the others almost entirely to the role of disinfectants—for use, that is, to do the best that can be done with table tops, walls, garbage cans, and anything too delicate for or inaccessible to steam under pressure. Some things are best incinerated, and ultraviolet light has limited usefulness.

The exception I mentioned is hexachlorophene, a phenol derivative with six atoms of chlorine in its molecule. (You may as well know that the "hex" is used here as it is in "hexagon" and has nothing to do with witches.) This substance, known since 1941, has acquired a nearly if not quite unique reputation as a slow but cumulatively effective agent against disease-producing staphylococci when it is used repeatedly on skin either in a liquid detergent mixture or incorporated into bar soap. Occasional people are sensitive to it, but for the most part it has been found entirely safe. There is much individual variation in its effects: some people just don't respond. It doesn't seem to have any consistent effect on normal skin bacteria, nor is there any evidence that it has any use in prevention or cure of acne. There may be other good antiseptics waiting to be discovered, but none has proved itself as yet.

Tincture of iodine has been thrown out as a home remedy because it does more harm than good. Both the iodine itself and the alcohol it is dissolved in damage the cells lining a cut—a fact only partly revealed by the stinging irritation it causes—and tend to delay rather than to promote healing. In medical practice a very dirty wound is cleaned not with iodine or other antiseptics but by a process to which the French word *débridement* is given—the cutting away of diseased and damaged tissue to leave clean healthy surfaces for healing. If the tissues are clean and healthy, nothing promotes healing more than keeping them so and leaving them alone. *Clean* is the right word, meaning free from gross foreign matter. Microbes likely to be present as native to the skin or other wounded part, and even most chance environmental contaminants, are nearly always mopped up and destroyed by the scavenging cells and other defensive substances in the wound. It is precisely these cells and substances that are damaged or knocked out by antiseptics, effects you may be unaware of unless exposed nerve endings are also irritated. The old myth that the stuff has to hurt to be good for you, like grandma's horrible-tasting tonic, is just that: an old myth.

Any attempt to help the tissues with chemical agents, except for antimicrobial drugs when there is a good reason for using them, is more than likely to do nothing but harm. This is true if the wound is anywhere on the skin, or in the mouth, or anywhere else. Even if it involves a perforation of the grossly microbe-laden intestinal tract, the body's normal defenses may be able to take care of the spillage. You will remember that the normal microbes do not usually initiate infection unless the tissues are abnormal, damaged by more than a simple wound, debilitated. If a drug is called for, it is an antimicrobial agent, not an antiseptic.

What is true of wounds is just as true of healthy skin and mucous membranes. Except for hexachlorophene in treatment or prevention of infection with pathogenic staphylococci, the use of antiseptics on the skin or any of the mucous membranes—the eye, the mouth, the nose and throat, the vagina, the rectum—is useless and likely to be harmful. Eyewashes, mouthwashes, and douches have no value even as cosmetics, despite certain opinions to the contrary even among professional people, who have merely picked these ideas uncritically out of the ancient lore to which we are all exposed. If washing is occasionally required, plain water serves as well as anything in the mouth, and table salt, a little less than 1 part in 100 parts of water, on other mucous membranes, including the eye. On skin, soap and water; and, if you really want the truth not too much soap, and not too often. As I have mentioned before, removing too much of the fatty material of the skin with an overzealous use of soap may be harmful. Know what you are doing to yourself, and let well enough alone.

But I seem to hear a question. I have my microbes, and you have yours. The fact seems to have got out that they are not precisely the same, mine and yours, since people talk of "mixing the breeds." True enough. And however much I may have learned to live at peace with my own microbes, does it follow that I could live just as peacefully with yours, and *vice versa*? The answer, if we are both healthy, is "yes." In any random crowd of people exchanging microbes, the answer is still "yes" if they are all healthy, but of course we can't count on that. The scientific study that provided the quotation at the head of this chapter illustrates the point. Even such relatively intimate contact as is entailed in passing a communion cup from mouth to mouth through the congregation is likely to be less dangerous than other things that inevitably happen at such times, especially the sneezing and coughing.

Again we must separate out questions of disease. If I am carrying pathogens in my respiratory tract, I may transmit them to you by various means of direct or indirect contact. Anyone excreting disease-producing microbes in feces or urine may infect you through the water you drink or swim in, or the food you eat. This happens especially among children by the more direct, commoner-than-you-may-think anus-to-finger-to-mouth sequence. Under certain special conditions a kiss can transmit syphilis; but with this and other venereal diseases the means of transit is almost exclusively sexual intercourse. It is an odd fact, and perhaps an encouraging one, that skin contact, with few exceptions apart from venereal disease, is relatively unimportant in the transmission of disease. Even staphylococci in pus must find their way to a wound and seem to prefer to settle on mucous membranes before they start growing in skin. Virus diseases with prominent skin symptoms, such as measles, German measles, chicken pox, and even smallpox, are transmitted in their early stages like respiratory diseases. As with tetanus and gas gangrene, fungus diseases are usually acquired by contact with soil or from other environmental sources. Malaria, yellow fever, plague—unless the last has become pneumonic—are transmitted only with the help of insects. This is a sketchy summary of rules that are not absolute; but they hold in general.

Many diseases make their way from one person to another via the respiratory tract. The list includes not only respiratory diseases proper, but others in which the microbe or virus is present in the nose or throat at some point, including mumps, meningitis, scarlet fever, and the measles-pox group. Sneezing and coughing get the infective material on its way, producing both large droplets that settle out fast and an aerosol or mist of much smaller particles that dry at once and tend to remain suspended. The larger droplets can carry the disease to people who happen to stand in their path, or the settled dust they may contaminate may be raised again and inhaled. Infected aerosols sometimes remain infective for long periods and get blown a long way; it is they, we have reason to think, that are responsible for major epidemics of what is called airborne disease. Dust is especially important as a vehicle with particular diseases, among them tuberculosis. Direct sneeze-to-face transmission is statistically least important. In this, if in no other way, it is equivalent to a kiss.

The subject of kissing belongs in this book, and I am glad to be able to speak well of it. It has that about it which can provoke the most liberal of bacteriologists to be either a puritan or a hypocrite. Yet the accumulated experience of the centuries has not condemned it; puritans in either black frocks or white coats have not been able to stop it; and by and large it seems that more damage may be done by the guilt that hypocrisy brings to it than by the microbes that use it as a bridge.

It happens that the mouth tends to remain free from the active diseases that affect the neighboring air passages, so that it does not harbor the massive infective accumulations found in the nose and throat. Nevertheless we know that the microbes and viruses of respiratory disease find their way into the mouth; and it is the mouth much more than the nose that provides the exit nozzle for both cough and sneeze. There is no doubt at all, in fact, that the whole respiratory group of diseases can be transmitted by kissing mouth-to-mouth. Indeed a few other infections need to be added, among them syphilis in its active secondary stage, herpes simplex (fever blisters) under particular circumstances only, and perhaps others, such as infectious mononucleosis, of which we know little.

We have no statistics with which to measure this risk. There is a risk—no doubt about it—but I suggest that it may be compared with the chance of being hit by a golf ball as you walk through the park. Even on the course the chance is small if you keep your eyes and ears open. Off the course, it falls to near zero. Enough of good practice based on germ theory has spread to most people. "Don't kiss me, I have a cold" is good manners as well as good sense. The Freudians are helping us to recognize that the mouth-to-mouth kiss is a sexual embrace and ought to be limited to lovers, with allowance for youthful experiments. You ought not to kiss the baby on the mouth. Not even your own baby.

Our normal microbes are entirely compatible with good health. The specter of infectious disease is fading now from the formidable aspect it had only a generation ago. But it still colors our outlook. The invisible germ remains as a vestige of our mythical fear of the dark. As science lights up one dark corner after another, the bigots who foisted Puritanism on us are retreating. But on the coattails of science comes another breed of fear-mongers, geniuses at turning uncertainties into nagging doubts and anxious fears, finding profit in every lingering patch of ignorance: the hucksters. We are so beset now with urgent advice, almost all of it false, that a manufacturer of wheat-germ oil, a food, takes full-page advertisements to protect his product. "What sort of germ did you say?" he seems to shout. "Please do not confuse my good germ with your bad ones."

Let us get back to the matter of mixing the breeds, speaking now of the normal microbes. We are exchanging them continually. We got them from other people in the first place, and we keep getting them and giving them. There is continual migration among the life on man in both directions. On any one of us, one group of microbes may displace another: a colony of one sort dies out or is destroyed by its neighbors, and a new one moves in to take its place. There is constant shifting, yet the principal microbic species and varieties remain the same. It all happens without our being in the least aware of it, unless, as we know, a new population happens to be actively pathogenic.

Yet our microbes are so much a part of us that, although we are capable of change, we tend to resist it. We do not easily or predictably accept new varieties of microbes. We—our defensive cells and fluids— join forces with our microbes in fending off the incursion of immigrants. We and the life on us seem to fight together to retain our individuality. Studies of precisely defined types among colon bacilli, a single species, have shown that in a given person's intestine one type may persist for months or even years and eventually be replaced by another, while additional transient types may come and go in the meantime, each remaining only a few days or weeks. In a middle-aged couple intensively examined, there was no interchange of types of colon bacilli. Efforts by the same experimenters to implant new types in dogs by a variety of means were all unsuccessful. The microbic population shifts, but more than the simple introduction of a new variety is needed to give it a foothold—something as yet unidentified. In a research on seventy-five married couples whose noses and throats were swabbed and cultured once a month for more than a year, in which the emphasis was on bacterial species in the borderland between normal and pathogenic, the most random sort of distribution turned up, with remarkably little evidence of interchange between spouses.

With exceptions to be mentioned that are both few and extreme enough to be called outrageous, no information has come my way during many years of searching for it to suggest that our normal microbes do our neighbors or our intimate companions any harm at all. This is true not only of the relatively acceptable microbes of mouth and skin, but also of those of feces. If pathogenic microbes or viruses are present, they can be transmitted from any of these sources, and only prejudice can make us think that one source is more or less menacing than another. But if only the normal microbes are there—even if they happen to be involved in disease and are accordingly present in enormously overgrown numbers—no harm accompanies their transit; the diseases in question, as I said before, are not communicated from person to person. Kissing your grandfather on the mouth may be distasteful to you if he has pyorrhea—you know you should have kissed him on the cheek!—but you won't get pyorrhea that way. You have the necessary microbes to start with, and the disease develops only if your mouth provides the opportunities I spoke about in Chapter 13.

The outrageous exceptions appear if you get bitten by somebody with pyorrhea or one of the related mouth diseases, or if you are reckless enough to punch such a person in the teeth. Injected this way in enormous numbers in lacerated skin, these microbes can start serious infections much like the ones we produce in the laboratory by injecting scrapings from the gums into guinea pigs. Contamination of a flesh wound with feces is also likely to be serious. The components of such events that make them exceptional and extreme are, first, the break in the surface of skin and damage to the tissues underneath, and second, the introduction of enormous numbers of microbes. It takes both together to make disease.

As we keep questions of disease in perspective, so we need also to keep questions of hygiene and sanitation within similar bounds. The purification and chlorination of drinking water have abolished cholera and massive epidemics of other intestinal diseases in this country. Sanitary plumbing that keeps sewers separate from the clean water supply helps hold Typhoid Mary in her place. We try to cure her, too, to get the typhoid bacilli out of her gall bladder; and meanwhile we try to keep her hands away from food prepared for others. Control of mosquito breeding and the use of screens have eliminated yellow fever from cities. Rat control, still shamefully imperfect, and the practice of bathing and changing our underwear at intervals to eliminate lice and fleas, have rid us of typhus fever and bubonic plague. Vaccines have worked wonders for a handful of diseases not easily dealt with by other means. Respiratory disease is still a major problem. Much has been accomplished by a kind of dilution—spacing out sweatshops and (planned for the week after next) slums. Ventilation and air-conditioning help indoors, and the sun disinfects whatever it can reach and penetrate. Antimicrobial drugs mop up some of the infection that remains.

The sun is a beneficent germicide, but its operative ultraviolet is easily stopped, and as an antiseptic it has the usual limitations. It cannot kill the microbes burrowing in your skin, let alone those in deeper crevices and orifices, before it burns, probably not until long after it burns. The layer of oil you use to keep from burning protects your microbes too.

All these hygienic and sanitary practices, which have made even the memory of the great plagues of the past begin to fade, have left the normal microbes unchanged, except for antimicrobial drugs, which have tended to upset them so as to make difficulties for us.

The plagues fade, but microbes remain; and vestiges of the memory of the formidable ones cling to those that live innocently, even beneficently, on man. Having done so much with hygiene and sanitation, we are beguiled into believing there is no limit to the good they can do for us. Once we worshiped Stercus; now he has joined Lucifer, and in his place we have erected a god of plumbing. The bathroom becomes clean, elegant, magnificent; even the toilet paper is delicately tinted, scented, and decorated with little patterns of flowers. Its very lavatory function is hidden under the great American roadside euphemism "rest room." Puritanism has all but disappeared, but hucksterism has moved in with new taboos, new fetishes, a new orthodoxy with a new motive. What is old is that we are being misled once again. We are being led into turning Freud's anal compulsion into a national neurosis.

Objective geographers ought to map the advertising jungle for us in detail. Years ago there were consumer groups that proposed to set forth bravely on this worthy enterprise; but they seem to have got sidetracked by expensive gadgetry. I can't do more here than sketch out part of the territory and suggest some of the approaches to it. Perhaps, with the help of other things I have talked about by way of general principles, you can do the rest yourself. It isn't easy. The huckster finds a seed of truth and nurtures it into a monstrous growth. Science is always incomplete, and he can find the holes in it and crawl through them.

I have no particular quarrel with cosmetics, and I don't intend to bother with patent medicines. More reluctantly I pass over the field of dentifrices and other tooth- and mouth-toiletries, except to suggest that they belong in the jungle with the rest. There are professional opinions to the contrary, I know. Some build on knowledge that tooth-brushing itself is desirable, a bare fact out of which nobody can make much money. Official approval of certain toothpastes seems to have been forced by research that has almost no relationship with the real world. Buy the simplest and cheapest one you can find: its main job is today what it always was, to make toothbrushing more agreeable. Brush your teeth to keep them clean, as you wash your face.

Let me say a few words more about matters I intend to leave out. Cosmetics, for instance. I have already voted for embellishment. I have opinions about dress, paint, artificial bulgers and flatteners, wigs and depilatories, packaged eyelashes and fingernails; but I shall keep them to myself. Within limits I may share your enthusiasm for perfume; but maybe we ought to stop at times to wonder why we like to smell like flowers or coconuts or little Asiatic deer or the guts of a sperm whale; couldn't we learn to love the smell of healthy men and women? It's only a step from the self-deception involved here to the attempt to keep yourself from growing old with the oils of ancient Egypt. Who is being fooled?

The ads are turning us all into compulsive hand-washers, and face- and body-washers too. We must dress elegantly and preferably beyond our budgets, precisely or with studied carelessness. Heads must be meticulously barbered, lacquered, every last hair in place, or, again, carefully schooled to look unschooled. Teeth of gods and goddesses;

no compromise with perfection; nails immaculately manicured or shaped and painted like talons. Dress and smell like aristocrats, look affluent whether you are or not; it is the image, not the man or woman beneath, that counts.

What Professor Kira calls "total cleanness" is equated with minimal social acceptability. We are all straightening the other fellow's tie, picking threads off his coat, flicking the last speck of dust off our own. If this is something the hippies are rebelling against, swinging all the way to the other side, are they perhaps trying to say something we ought to listen to?

As Freud tells us, the compulsively neat anal person, underneath the façade he so anxiously offers to the world, is horribly, hopelessly dirty, fighting in a sea of filth for a decontamination he can never achieve. This is what the advertising men seem to want us all to be.

Your face, the magazine page says, is loaded with invisible dirt . . . a magnet for dirt and germs . . . rich or poor, colonel's lady or Judy O'Grady, you'd be ashamed if you knew. Are you afraid to let him see your skin close up? Does it begin to show the unforgivable lines of premature old age? Of all the ugly parts of you that you try to hide, aren't your feet the ugliest? There's a remedy for each of these evils. If one doesn't work or does more harm than good, there's another, and a remedy for the remedy, and so on *ad nauseum.* If all else fails, there's paint on a stick to cover everything over.

Most of this drivel is aimed at girls and women, but as men become more colorful, bejeweled, and perfumed—and who will dispute that sex equality ought to work both ways?—they in their turn are being whispered to about offending body odors and the ominous presentiments of dwindling masculinity. But I leave the men to fend for themselves, and invite you rather to see what is happening to the girls, who are taking the brunt of the bombardment. Let's look particularly at the huckster who plants and nurtures anxiety and fear in the minds of adolescent girls, new brides, and young mothers. Could this be the fellow Falstaff had in mind when he said no hole in hell were hot enough for him?

The girls have a fearsome lot of problems all their own, the advertisements say—all, of course, the work of microbes, or germs as the huckster likes to call them, those nasty little things which, he is sure you know, need to be exterminated, like vermin, trod on like cockroaches on the kitchen floor. See where they are stirring up offensive smells to affront fastidious noses, especially the incredibly sensitive olfactory organs of men. Unless you heed the huckster's friendly counsel you are sure to give off these unpleasant if not loathsome stinks from your skin and your mouth; but far worse is something especially abhorrent which he delicately calls "feminine odor" and then localizes and details for you.

The problems begin in the baby's wet diaper, where ammonia the germs make out of urine is the cruel world's substitute for the protective vernix that covers the baby at birth. A mystic lotion neutralizes the ammonia and kills the nasty germs. If you are a wise mother, the benevolent words whisper, you will not deny your infant this nostrum—and some others the same alchemist has brewed for you—but will apply it regularly. Otherwise don't expect your little girl to be beautiful when she gets bigger; surely you are not naïve enough to trust nature with the job! (Little boys, unmentioned, are presumably immune—or maybe their beauty doesn't matter.) But watch out! Here is a young mother, a serenely victorious smile on her face, squirting something from an aerosol can to clear the air and make it salubrious. I hope the can says on it somewhere, "Don't spray the baby!" Keep it away from bottle, teething rings, toys. It ought to tell you to keep it in the cellar, use it for vermin and mildew. (Another genius advises you to spray your telephone—with something else—since "phone germs" transmit colds and goodness knows what not.) We may expect to see the germ-free isolator soon, the adman's fantasy of a baby inside, handled only with rubber gloves, cleaner than the prematures in the hospital nursery. Total cleanness! Poor baby!

The sophisticated mother having protected her child with all necessary drugstore armor, the young thing arrives at adolescence and begins to find her own way through the advertising jungle. Skin oily? Later on it will be too dry, but now you must scrub and soap the oil away, medicate the devilish germs. Otherwise, no boys! You don't really think a boy will come near you with that pimple, do you? Some of the magical potions purveyed with these frightening suggestions actually promise, or seem to promise, that they can prevent or cure acne. Usually they are just "for" girls with "blemishes": they "fight" or "combat" bacteria, which sounds encouraging. They help control something not quite specified on a teenager's broken-out skin. But sometimes they kill germs by the million, and some actually "help to prevent and control" infection of pimples. Anyway the message comes through, and the Food and Drug Administration seems to be helpless. One cure-all even dries up oil, kills bacteria, reduces inflammation, clears up blemishes, and, if that isn't enough, prevents new acne pimples too. What more can you ask for? Still another, proclaiming that it contains hexachlorophene, puts the hex on germs, peels away dead skin, and contrives with other evident magic to prevent acne from ruining your fun. How wretched a young girl's life would be without these kind people!

A few years pass. The blemishes have gone away. The boys have appeared, and our spotlessly beautiful young Venus has found herself a mate, presumably equally immaculate. But the new bride faces new perils in the ugly germ-laden world. Only the wisdom of the adman, this time perhaps with the same sort of humane female collaboration, can overcome the germs that now threaten the marriage at its very consummation. There is a special feminine odor, its loathsomeness unspecified but not to be questioned. Scanning the fine print in the advertisements, young eyes in fear and trembling learn that it is in the "outer vaginal" area that this abomination is to be nosed out (what an unutterably horrid idea!). The context allows no reassuring doubt. The smell is inside too. It comes from the curse of menstruation, abetted by that other evil always delicately called perspiration; both, of course, being meddled with, befouled, by those despicable germs. It is not to be imagined, not for a moment, that anything so mundane (and cheap) as soap and water could make the immaculate bride ready for the arms of her lover. But take heart. Here are douches, sprays, powders, suppositories in bountiful array, each one more promising than the last. Science is ready to rescue the maiden in her final maidenly distress. Benevolent antiseptics, deodorants, anti-perspirants, especially designed to substitute for careless nature, saviors of femininity. Use them at bedtime, and get used to them. Later on they will be indispensable for intimate hygiene during the day, at work, to protect you from offending the crowds around you through the long hot hours.

According to the huckster, if the young mother fails to purify the air of the nursery, if the adolescent girl doesn't "hex" those pimples, if the new bride disdains to sterilize her perineum, they are all offending against decency. That means, if I have been making any sense, that they are all obscene.

Obscenity is like the shuffled pea in the old shell game; you get fooled every time. The world is full of it, but the quick hand directs your eye to the wrong shell. Is it you who "offend" or the adman who offends against you? Is it the healthy body, its parts, functions, and products, or the exhalations of automobiles and smokestacks? Which shell do you choose, words or deeds? Is it the "obscenities" hurled by unarmed civilians, or the swinging nightsticks and billowing nausea gas of helmeted and masked police? Is it normal microbes or perverted men?

Notes
Bibliography
Index

Notes

Chapter 2: Leeuwenhoek Saw It First

PAGE

14. *Pissabed.* Dobell identifies this as the woodlouse or sowbug, *Oniscus asellus.* Partridge (*Dictionary*) defines *piss-a-bed* as dandelion, which is the meaning of the equivalent word in French (*pissenlit*), and notes that the plant has diuretic properties. *The Merck Index of Chemicals and Drugs* (1960) lists two compounds derived from the dandelion (Taraxacum and Leontodin). "Lion's tooth" presumably refers to the leaf.

18. *recent estimate* (bacteria of the mouth). See Dale; Socransky.

22. *Accurate counts* . . . [of saliva]. See Rosebury (*Microorganisms*), p. 314. Additional references for this chapter are also given in this source.

Chapter 3: The Facts of Life on Earth

26. *the Russian biochemist A. I. Oparin.* The whole subject of the origin of life has been admirably reviewed by Bernal, who also gives, as appendices, Oparin's original paper, dated 1924, translated into English for the first time, and Haldane's 1929 paper, both in full. In this original paper, Oparin says explicitly that " 'autotrophic' feeding" could only have evolved after a simpler and more primitive type of life had emerged. The idea that autotrophic bacteria were among the first living things was put forward in 1936 by the English biochemist Knight.

Chapter 4: This Is the Life

31. *bacteria . . . in feces.* See Rosebury (*Microorganisms*).
37. *more recent experiments with mice.* See Dubos; Savage.

215

37. *upper region of the intestine.* See Gorbach.
the microbes keep coming. See Gibbs.
39. *the navel, etc.* See Riely.

Chapter 5: At Home on Man

43. (Chapter 5.) For sources see Rosebury (*Microorganisms*).
47. *Albert Finney and Joyce Redman.* Tom and Jenny, respectively, in the Osborne-Richardson movie *Tom Jones* (see Chapter 1).
51. *Germ-free animals.* For sources more recent than those given in Rosebury (*Microorganisms*) see Dolowy; Gordon; Schaffer.
53. *Milk cattle.* See Virtanen.

Chapter 6: Freud's Gold

64. *the sense of smell* (quotation 1, line 2). This is discussed by Havelock Ellis, who says, among other things, that girls and women tend to respond more strongly than do boys and men; that the odors concerned in sexual allurement (or repulsion) are evidently those that result directly from microbic action ("the armpit . . . [is] the chief focus of odor in the body"); and that perfume was used originally not to disguise but to fortify natural odors—hence the use of animal scents like musk and ambergris, and the persistence of attractive association for the scent of leather. Kinsey mentions smell only briefly, together with sight, taste, and hearing as means of erotic stimulation, and doubts that male and female responses are different.

Chapter 7: How Did We Go Wrong?

67. *recent studies* (of paleolithic art). See Leroi-Gourhan.
71. *witch burnings.* See Hughes.

Chapter 8: Don't Mention It

78. *cursing and swearing.* Ashley Montagu's book on this subject appeared while I was writing this one; and, having since read it with interest and enjoyment, I am not disposed to alter what I have said. Montagu's focus is sharply on malediction as an art form. He acknowledges Graves's *Lars Porsena* as his starting point, and includes, among other things, a scholarly and spirited defense of forbidden "four-letter

words," as well as an extensive collection of shorter pieces of artistic invective, both profane and polite. An earlier collection of longer examples of the latter species was made by Kingsmill.

78. *words.* I have emphasized the shorter words in common use. Among others, unlike the root word "dirt," some are always pejorative, beginning, perhaps, with *foul* and *filthy.* Consider *scurrilous, squalid, scabrous.* See Partridge (*Dictionary*) both for words defined and words used in definitions.

79. *emphatic old English words.* This quality of the permitted words also applies to the others, of course. The short "obscene" words are much more evocative than the longer classical ones.

80. *snobbery.* For a different slant see Mitford, on "U" or (British) upper-class speech and its "non-U" opposite.

85. *not the words themselves.* The note above (p. 78) applies here as well.

86. *sir-reverence.* Defined by Partridge (*Dictionary*) as "Human excrement; a lump thereof"; and in the same author's *Shakespeare's Bawdy* as "a catch-phrase uttered when one comes upon a lump of (human) excrement; the lump itself. . . ." But "save your reverence," the source phrase, is merely an apology for an unseemly expression or incident.

88. *explosively funny.* See the discussion of jokes in Chapter 14. I have noticed a different kind of laughter, which Bruce's performances may also have elicited, growing out of audience tension in avant-garde theater-of-the-absurd or alienation plays, in situations that did not seem at all humorous to me—especially among members of the audience very close to the actors in theater-in-the-round performances.

88. *Mark Twain.* There are hints of this philosophy, of course, in the "funny" books, notably *A Connecticut Yankee at King Arthur's Court* and *The Man That Corrupted Hadleyburg*; but the whole picture comes out only in the posthumous works: in his *Autobiography*, especially the earlier version (ed. A. B. Paine, P. F. Collier, 1925), in *Mark Twain in Eruption* (ed. B. DeVoto, Harper, 1922), and more recently in *Letters from the Earth* (ed. DeVoto, Harper & Row, 1962) and *The Damned Human Race* (ed. J. Smith, Hill & Wang, 1962).

Chapter 9: Toilet Training

96. *sophisticated water pipes.* Some of the material in this section is taken from Forbes.

99. *Dryden.* The point is exemplified by comparing Dryden's verse

with two recent translations of Juvenal, by H. Creekmore (Mentor, 1963) and P. Green (Penguin, 1967).

100. *bidet.* While working on this book I have been noticing an ad that has appeared repeatedly in *The New York Times* for a bidet-like device to be substituted for the usual toilet seat, with a built-in water jet and a drying stream of warm air. It costs between $230 and $240, installed and serviced, and is hence evidently not intended for the general run of people. We are told that it is "the most wonderful advance in bathroom habits since indoor plumbing itself."

104. *graffiti.* There is currently a little magazine by this name, and I have noticed at least one other collection.

105. *Byron.* This is one of the poet's *Ephemeral Verses* as given in the Cambridge Edition of his complete works (Houghton Mifflin Co., Boston, 1933, p. 235). I have not been able to trace the lines on p. 112.

Chapter 10: *The Romans Had a God for It*

110. *Baal-Peor.* Various lexicons dating between 1892 and 1911 that I have consulted (and, because of the dates, among other things, on whose complete credibility I reserve judgment), give the sense of *Baal* or *Bel* as a lord, god, or deity at perhaps a low level, and of *Peor* as an opening, cleft, or hiatus, but consistently sexual rather than anal. *Phegor*, given as *Phagor* and in other spellings, seems to be Greek rather than Hebrew; and I have been able to find no support for Bourke's suggestion of a relationship to Crepitus.

114. *Cockledy bread.* A similar verse that mentions "cockelly-bread," and another almost identical with the second one in the text and credited to the same sources Bourke used, appear in a long footnote in *The Old Wives' Tale* by George Peele, at the end of a verse in the text spoken by a Head that rises in the Well of Life:

> Gently dip, but not too deep,
> For fear you make the golden beard to weep.
> Fair maiden, white and red,
> Stroke me smooth, and comb my head,
> And thou shalt have some cockell-bread.

The footnote, presumably supplied by Peele's editor, F. B. Gummere (1903), begins, "After many inquiries on the important subject of *cockell-bread,* I regret to say I am unable to inform the reader what is was." And in fact what follows throws little further light on the subject even when its contained Latin is translated, except to suggest that "cockle" may be an antiquated Norman word for "buttocks" and that cockle-bread was a love-philter. A possible connection with the French *coquille* (shell), which in English becomes "cockle-shell," is mentioned but discounted. Partridge seems to have overlooked this "important subject." I was led to Peele when I ran across the lines just quoted in an anthology of Elizabethan verse (Penguin, 1965), to which the editor, Edward Lucie-Smith, appends the otherwise unattributed note, "Cockle-bread is a magical cake, used as a love-charm, and made with menstrual blood." It is unlike Penguin editors to invent such ideas, but I have no way of knowing where this one came from.

118. *early Christian practices.* For a connection between such practices and the later persecution of witches, see Hughes.

119. *Arabian Nights.* Burton gives additional details in two footnotes to the word "skite" and after "ten countries," both in the first passage quoted. The first of these may be given in full:

> Again the coarsest word "Khara." The allusion is to the vulgar saying, "Thou eatest skite!" (*i.e.*, thou talkest nonsense). Decent English writers modify this to, "Thou eatest dirt": and Lord Beaconsfield made it ridiculous by turning it into "eating *sand.*"

In the second footnote Burton says, "the idea of the Holy Merde might have been suggested by the Hindus," and speaks of a prelate "carrying ox-dung and urine to the King, who therewith anoints his brow and breast, &c. . . ."

Chapter 11: Husbandry on Earth

130. *nothing whatever on this subject.* A paper by Theodor Wieland, "Poisonous principles of mushrooms of the genus Amanita," (*Science* 159:946–952, 1968) also says nothing of urine-drinking; but a more recent paper by Norman R. Farnsworth, "Hallucinogenic plants," in the same journal (162:1086–1092, 1968) not only mentions the subject, giving Mr. Wasson as source (see p. 133), but gives the additional information from other sources that the principal "psychotropic" constituent of *Amanita muscaria* is not muscarine but ibotenic acid, one of six substances other than muscarine that have been detected in the fly agaric. See also R. E. Schultes, "Hallucinogens of plant origin" (*Science* 163:245–254, 1969).

137. *"flap-dragon."* Partridge (*Dictionary*) gives, for the sense here intended, only "a raisin snatched from burning brandy and eaten hot." This appears to be the sense in which Shakespeare used the term in *Love's Labour's Lost* (V, i, 43), *Henry IV*, Part 2 (II, iv, 244), and

A Winter's Tale (III, iii, 100). But Partridge notes that it was used after Shakespeare's time to mean syphilis or gonorrhea (the two diseases were not distinguished until 1832), and otherwise colloquially as a pejorative.

Chapter 12: The Great Scatologists

150. *Rabelais.* The quotations on this and the following pages are from Putnam

154. *Shakespeare.* See Partridge, *Shakespeare's Bawdy.*

155. *censorship.* See Craig; Hyde.

156. *as recently as 1933.* The quotation is from C. A. Moore.

165. *posterior trumpeter.* Partridge (*Shakespeare*, p. 11) speaks with disfavor of the "days when, as at the end of the 17th Century, a pamphlet dealing with a noisy venting and written by a pseudonymous Don Fartando could be published and enjoyed and when the ability to play tunes by skilfully regulating and controlling one's windy expressions was regarded as evidence of a most joyous and praiseworthy form of wit." But much more recently one Joseph Pujol, whose virtuosity in this art form was confirmed and explained by physicians who examined him, enjoyed a considerable vogue as an entertainer in France and Belgium; and a book about him seems to have been popular still more recently in England. See Nohain. Douglas also speaks of this phenomenon but with a suggestion that it may have entailed cheating. The evidence in the source cited convinces me that M. Pujol was a man of the greatest integrity.

Chapter 13: Trouble

167. (Chapter 13.) References for much of the material on endogenous infection are given in Rosebury (in Dubos and Hirsch).

167. *Hoist with mine own petard.* The last word in this paraphrase of Hamlet (III, iv, 208) can hardly be allowed to pass without noting its derivation, best given by Eric Partridge in his *Origins, a Short Etymological Dictionary of Modern English* (Routledge and Kegan Paul, London, 2nd ed., 1959, revised 1966) as from the early modern French *pétard*, detonator, literally "farter," from the Middle French *peter* (later *péter*), to explode, originally to break wind. The word came to mean, in Shakespeare's time, a kind of artillery to batter, or break down doors; but we need not doubt that the bard knew its French meaning, and used it for the color of this metaphorical sense.

175. *acne vulgaris.* See Crounse; Hicks; Mullins; and Savin.

175. *two Philadelphia dermatologists.* See Kirschbaum.

176. *increased bacterial growth on the skin.* See Lacey.

177. *chronic respiratory diseases.* See Anderson; Cherniak; David; and Tomashefski. A recent issue of *The Yale Journal of Biology and Medicine* (Vol. 40, April–June, 1968) is devoted to these problems.

178. *"potentially contaminated" abdominal operations.* See Bernard.

178. *urinary tract infections.* See Allen; Dontas; Hinman; LeBlanc; Norden; and Stansfield.

179. *kwashiorkor.* See Lehndorff; Phillips and Wharton; Salomon; and Smythe. Evidence was given in a CBS telecast I saw on May 21, 1968, that this disease is present among the Navajo Indians in this country. Confirmation appeared in the June 28, 1968, issue of *Science* (Bryce Nelson, v. 160, p. 1435).

179. *noma.* See Dawson; Jeliffe. The quotation on p. 180 is from the latter.

181. *Pyorrhea.* Details and references are in Rosebury (in Dubos and Hirsch).

Chapter 14: Obscenity Reconsidered

189. *Artistic invective.* See the note to p. 78.

190. *Freud.* See *Jokes and Their Relation to the Unconscious*; and the article "Humour" in *Complete Psychological Works*.

191. *a psychiatrist.* See Hartogs.

194. *without any hypocrisy.* Something like this has been done by A. S. Neill at Summerhill School in Leiston, Suffolk, about 100 miles from London (see Neill). But science does not seem to be given much emphasis in the teaching there, and microbiology may not even be mentioned. General sex education, however, is evidently handled with exemplary honesty and freedom. The work of this school has attracted much favorable notice, and some hostility.

Chapter 15: The Shell Game

200. These "pyocyaneus" microbes—more properly *Pseudomonas aeruginosa*—have actually been found contaminating hospital *disinfectants*! See Burdon. These and other microbes have also been observed lurking in hand lotions used in hospitals. See Morse. Some of the literature more recent than that given in my *Microorganisms* . . . can be found in Jellard and in Phillips.

200. *Sterilization of skin.* See King; Lowbury; Speers.

201. *Hexachlorophene.* See the references under sterilization of skin (p. 200); also Gezon; Gluck; Macpherson.

203. *skin contact.* The relative importance of disease transmitted from one skin surface to another—apart from venereal disease—needs the qualification, *except in the tropics,* and in particular *except where children go naked.* Among diseases that tend to be spread among such children is yaws, a widespread, serious, non-venereal infection caused by a spirochete almost identical with that of syphilis.

205. *a middle-aged couple.* See Sears.

206. *seventy-five married couples.* See Harvey.

207. *antimicrobial drugs, which have tended to upset* (the normal microbes). This subject has been reviewed by Weinstein.

Bibliography

Allen, T. D. "Pathogenesis of urinary-tract infections in children." *New England Journal of Medicine* 273:1472–1477, 1965.

Anderson, D. O., and Ferris, B. G. "Role of tobacco smoking in the causation of chronic respiratory disease." Ibid. 267:787–794, 1962.

Barnes, H. E. *An Intellectual and Cultural History of the Western World*, 3rd ed., 3 vols. New York: Dover Publishing Co., 1963.

Bernal, J. D. *The Origin of Life*. Cleveland: World Publishing Co., 1967.

Bernard, H. R., and Cole, W. R. "Wound infections following potentially contaminated operations." *Journal of the American Medical Association* 184:290–292, 1963.

Bogoras, W. "The Chukchee." *Memoirs of the American Museum of Natural History* 11:205–207, 1904.

Bourke, J. G. *Scatologic Rites of All Nations*. Washington, D.C.: W. H. Lowdermilk & Co., 1891.

Brill, A. A. *The Basic Writings of Sigmund Freud*. New York: Modern Library, 1938.

Brown, N. O. *Life against Death. The Psychoanalytic Meaning of History*. New York: Vintage Books (Random House), 1959.

Bruce, L. *How to Talk Dirty and Influence People*. Chicago and New York: Playboy Press, and Pocket Books (Simon & Schuster), 1963. See also Cohen, J. *The Essential Lenny Bruce*. New York: Ballantine Books, 1967.

Burdon, D. W., and Whitby, J. L. "Contamination of hospital disinfectants with *Pseudomonas* species." *British Medical Journal* 2:153–155, 1967.

Calvin, M. "Chemical evolution and the origin of life." *American Scientist* 44:248–263, 1956.

Cherniack, N. S. and others. "The role of acute lower respiratory infection in causing pulmonary insufficiency and bronchiectasis." *Annals of Internal Medicine* 66:489–497, 1967.

Craig, A. *Suppressed Books: A History of the Conception of Literary Obscenity*. Cleveland and New York: World Publishing Co., 1966.

Crounse, R. F. "The response of acne to placebos and antibiotics." *Journal of the American Medical Association* 193:906–910, 1965.

Dale, A. C., and others. "Quantitative determination of bacteria in the gingival crevice of man." *Journal of Dental Research* 40:716, 1961.

David, W. "Chronic respiratory disease—the new look in the Public Health Service." *American Journal of Public Health* 57:1357–1362, 1967.

Dawson, J. "Cancrum oris." *British Dental Journal* 79:151–156, 1945.

Dobell, C. *Antony van Leeuwenhoek and His "Little Animals."* New York: Dover Publishing Co., 1960.

Dolowy, W. C., and Muldoon, R. L. "Studies of germfree animals I. Response of mice to infection with influenza A virus." *Proceedings of the Society for Experimental Biology and Medicine* 116:365–371, 1964.

Dontas, A. S., and others. "Bacteriuria in old age." *Lancet* 2:305–316, 1966.

Douglas, N. *Some Limericks*. New York: Grove Press, 1967.

Dubos, R. J., and others. "Indigenous, normal, and autochthonous flora of the gastrointestinal tract." *Journal of Experimental Medicine* 122:67–76, 1965.

Ellis, Havelock. *Psychology of Sex*. New York: Emerson Books, 1933.

Forbes, R. J. *Studies in Ancient Technology*, vol. 1. Leiden: E. J. Briel, 1955.

Ford, W. W. *Transactions of the Association of American Physicians* 38:225. Cited by Grossman, C. M., and Malbun, B. "Mushroom poisoning." *Annals of Internal Medicine* 40:249–259, 1954.

Frazer, Sir J. G. *The Golden Bough*. New York: Macmillan, 1947.

Freud, Sigmund. *A General Introduction to Psychoanalysis*. Garden City, N.Y.: Garden City Publishing Co., 1938.

———. *Complete Psychological Works*. London: Hogarth Press, 1950.

———. *Jokes and Their Relation to the Unconscious*. New York: W. W. Norton & Co., 1963.

Gibbs, B. M., and Stuttard, L. W. "Evaluation of skin germicides." *Journal of Applied Bacteriology* 30:66–77, 1967.

Gogarty, O. St. J. *As I Was Going Down Sackville Street: A Phantasy in Fact*. New York: Reynal & Hitchcock, 1937.

Goldsmith, O. *The Citizen of the World; or, Letters from a Chinese Philosopher Residing in London*, no. XXXII. London: J. Newberry, 1762.

Gorbach, S. L., and others. "Studies of intestinal microflora II. Microorganisms of the small intestine and their relations to oral and fecal flora." *Gastroenterology* 53:856–867, 1967.

Gordon, H. A., Wostmann, B. S., and Bruckner-Kardoss, E. "Effects of microbial flora on cardiac output and other elements of blood-circulation." *Proceedings of the Society for Experimental Biology and Medicine* 114:301–304, 1963.

Graves, R. *Lars Porsena, or the Future of Swearing and Improper Language*. London: Kegan Paul, Trench, Trubner & Co., 1927.

Haldane, J. B. S. "The Origin of Life." *Rationalist Annual*, 1929. (Also in *Science and Human Life*, 1933; and see note to p. 26.)

Harington, Sir John. *The Metamorphosis of Ajax, 1596*. A Critical Annotated Edition by E. S. Donno. New York: Columbia University Press, 1962.

Hartogs, R. (with H. Fantel) *Four-Letter Word Games: The Psychology of Obscenity*. New York: M. Evans & Co. (Delacorte Press), 1967.

Harvey, H. S., and Dunlap, M. B. "Upper respiratory flora of husbands and wives. A comparison." *New England Journal of Medicine* 262:967–977, 1960.

Heer, F. *The Intellectual History of Europe*, 2 vols. Garden City, N.Y.: Doubleday Anchor Books, 1968.

Herrick, R. *The Complete Poetry of Robert Herrick*. Edited with an Introduction and Notes by J. Max Patrick. New York: 'V. W. Norton & Co., 1968.

Hicks, J. H. "Demethylchlortetracycline: a double-blind study in the treatment of acne with attention to side-effects noted." *Southern Medical Journal* 55:357–360, 1962.

Hinman, F., Jr. "Mechanisms for the entry of bacteria and the establishment of urinary infection in female children." *Journal of Urology* 96:546–550, 1966.

Hobbs, B. C., Knowlden, J. A., and White, A. "Experiments on the communion cup," *Journal of Hygiene* 63:37–48, 1967.

Hughes, P. *Witchcraft*. Baltimore: Penguin Books, 1965.

Hyde, H. M. *A History of Pornography*. New York: Dell Books, 1966.

Jeliffe, D. B. "Antibiotic treatment of infective gangrene of the mouth." *Journal of Tropical Medicine and Hygiene* 56:53–56, 1953.

Jellard, C. H., and Churcher, G. M. "*Pseudomonas aeruginosa* (*pyocyanea*) infection in a premature baby unit, with observations on the intestinal carriage of *Pseudomonas aeruginosa* in the newborn." *Journal of Hygiene* 65:219–222, 1967.

Jones, F. A. "Burbulence: a fresh look at flatulent dyspepsia." *Practitioner* 198:367–370, 1967.

Kennan, G. *Tent Life in Siberia*. New York: G. P. Putnam's Sons, 1910.

King, T. C., and Zimmerman, J. M. "Skin degerming practices: chaos and confusion." *American Journal of Surgery* 109:695–697, 1965.

Kingsmill, H. *An Anthology of Invective and Abuse* (1929); *More Invective* (1930). New York: Dial Press.

Kinsey, A. C., Pomeroy, W. B., Martin, C. E., and Gebhard, P. H. *Sexual Behavior in the Human Female*. Philadelphia: W. B. Saunders Co., 1953.

Kira, A. *The Bath Room: Criteria for Design*. New York: Bantam Books, 1966.

Kirschbaum, J. O., and Kligman, A. M. "The pathogenic role of Corynebacterium acnes in acne vulgaris." *Archives of Dermatology* 88:832–833, 1963.

Knight, B. C. J. G. "Bacterial nutrition; materials for a comparative physiology of bacteria." *Medical Research Council Special Report Series No. 210*, 1936.

Krassner, P. (Obituary of Lenny Bruce.) *Ramparts*, pp. 37–38, October 1966.

Lacey, R. W. "Antibacterial action of human skin. *In vivo* effect of acetone, alcohol and soap on behavior of *Staphylococcus aureus*." *British Journal of Experimental Pathology* 49:209–215, 1968.

LeBlanc, A. L., and McGanity, W. J. "The impact of bacteriuria in pregnancy—a survey of 1300 pregnant patients." *Texas Reports on Biology and Medicine* 22:336–347, 1964.

Lehndorff, H. "Kwashiorkor: Present problems." *Archives of Pediatrics* 78:293–316, 1961.

Leroi-Gourhan, A. "The evolution of paleolithic art." *Scientific American*, pp. 59–70, February 1968.

Lowbury, E. J. L., Lilly, E. A., and Bull, J. P. "Methods for disinfection of hands and operation sites." *British Medical Journal* 2:531–536, 1964.

Miller, S. L. "A production of amino acids under possible primitive earth conditions." *Science* 117:528–530, 1953.

Miner, H. "Body ritual among the Nacirema." *American Anthropologist* 58:503–507, 1956.

Mitford, N. (editor). *Noblesse Oblige*. New York: Harper & Brothers, 1956.

Montagu, A. *The Anatomy of Swearing*. New York: Macmillan, 1967.

Moore, C. A. Introduction to *Twelve Famous Plays of the Restoration and Eighteenth Century*. New York: Modern Library, 1933.

Morse, L. J., and Schonbeck, L. E. "Hand lotions—a potential nosocomial hazard." *New England Journal of Medicine* 278:364–369, 1968.

Mullins, J. F., and Naylor, D. "Glucose and the acne diathesis: an hypothesis and review of pertinent literature." *Texas Reports on Biology and Medicine* 20:161–175, 1962.

Neill, A. S. *Summerhill: A Radical Approach to Child Rearing*. New York: Hart Publishing Co., 1960.

———. *Freedom—Not License!* New York: Hart Publishing Co., 1966.

Nohain, J., and Caradec, F. *Le Petomane, 1857–1945*. Los Angeles: Sherbourne Press, 1968.

Norden, C. W., and Kass, E. H. "Bacteriuria of pregnancy—a critical appraisal." *Annual Review of Medicine* 19:431–470, 1968.

Oparin, A. I. *The Origin of Life on the Earth*. New York: Macmillan Co., 1938. (See note to p. 26.)

Partridge, E. *A Dictionary of Slang and Unconventional English*, 6th edition. New York: Macmillan Co., 1967.

———. *Shakespeare's Bawdy*. New York: E. P. Dutton & Co., 1960.

Peele, George. "The Old Wives' Tale," in *The Minor Elizabethan Drama*, vol. 2, *Pre-Shakespearean Comedies*. London: J. M. Dent (Everyman edition), 1910, footnote 2, pp. 150–151.

Phillips, I. "*Pseudomonas aeruginosa* respiratory tract infections in patients receiving mechanical ventilation." *Journal of Hygiene* 65:229–235, 1967.

————. and Wharton, B. "Acute bacterial infection in kwashiorkor and marasmus." *British Medical Journal* 1:407–409, 1968.

Putnam, S. *All the Extant Works of François Rabelais*, 3 vols. New York: Covici-Friede, 1929.

Reynolds, R. *Cleanliness and Godliness*. New York: Doubleday & Co., 1946.

Riely, R. E., and Shorenstein, D. J. "Microbiological flora of human subjects under simulated space environments." Joint NASA/USAF Study AMRL-TR-66-171, October 1966.

Rosebury, T. "Bacteria indigenous to man." Chapter 14, pp. 326–355, in Dubos, R. J., and Hirsch, J. G. (editors), *Bacterial and Mycotic Infections of Man*, 4th ed. Philadelphia: J. B. Lippincott Co., 1965.

————. *Microorganisms Indigenous to Man*. New York: McGraw-Hill Book Co., 1962.

Salomon, J. B., Mata, L. J., and Gordon, J. E. "Malnutrition and the common communicable diseases of childhood in rural Guatemala." *American Journal of Public Health* 58:505–516, 1968.

Savage, D. C., Dubos, R. J., and Schaedler, R. W. "The gastrointestinal epithelium and its autochthonous bacterial flora." *Journal of Experimental Medicine* 127:67–76, 1968.

Savin, R. C. "Antibiotics and the placebo reaction in acne." *Journal of the American Medical Association* 196:365–367, 1966.

Schaffer, J., and others. "Response of germ-free animals to experimental virus monocontamination I. Observation on Coxsackie B virus." *Proceedings of the Society for Experimental Biology and Medicine* 112:561–564, 1963.

————. "Studies on fatal hypoglycemia in axenic (germfree) piglets." Ibid. 118:566–570, 1965.

Schierbeek, A. *Measuring the Invisible World: The Life and Works of Antoni van Leeuwenhoek FRS*. London and New York: Abelard-Schuman, 1959.

Sears, H. J., and others. "Persistence of individual strains of *Escherichia coli* in man and dog under varying conditions." *Journal of Bacteriology* 71:370–372, 1956.

Smythe, P. M. "Changes in intestinal bacterial flora and role of infection in kwashiorkor." *Lancet* 2:724–727, 1958.

Socransky, S. S., and others. "The microbiota of the gingival crevice area of man I. Total microscopic and viable counts and counts of specific organisms." *Archives of Oral Biology* 8:275–280, 1963.

Speers, R., Jr. and others. "Increased dispersal of skin bacteria into the air after shower baths: the effect of hexachlorophene." *Lancet* 1:1298–1299, 1966.

Stansfield, J. M. "Clinical observations relating to incidence and aetiology of urinary-tract infections in children." *British Medical Journal* 1:631–635, 1966.

Taylor, N. *Narcotics: Nature's Dangerous Gifts.* New York: Delta (Dell Publishing Co.), 1963.

Tissier, H. *Récherches sur la Flore Intestinale des Nourissons (État Normal et Pathologique).* Thesis, Faculty of Medicine of Paris. Paris: Georges Carré et C. Naud, Editeurs, 1900.

Tomashefski, J. F., and Pratt, P. C. "Pulmonary emphysema: pathology and pathogenesis." *Medical Clinics of North America* 51:269–281, 1967.

Tynan, K. Foreword to Bruce (see above), 1963.

Virtanen, A. I. "Milk production of cows on protein-free feed." *Science* 153:1603–1614, 1966.

Wasson, R. G. "Seeking the magic mushroom." *Life* v. 42, May 13, 1957.

Weinstein, L. "Superinfection: a complication of antimicrobial therapy and prophylaxis." *American Journal of Surgery* 107:704–709, 1964.

Wright, L. *Clean and Decent: The Fascinating History of the Bathroom and the Water Closet.* New York: The Viking Press, 1960.

Index